Spurious Coin

SUNY Series
Studies in Scientific and Technical Communication
James P. Zappen, editor

Spurious Coin

A History of Science, Management, and Technical Writing

Bernadette Longo

STATE UNIVERSITY OF NEW YORK PRESS

Published by
State University of New York Press, Albany

© 2000 State University of New York

Printed in the United States of America

For information, address the State University of New York Press,
State University Plaza, Albany, NY 12246

Production by Michael Haggett
Marketing by Anne M. Valentine

Library of Congress Cataloging-in-Publication Data

Longo, Bernadette, 1949–
    Spurious coin: a history of science, management, and technical
writing/ Bernadette Longo.
        p.   cm.—(SUNY series, Studies in scientific and technical
communication)
    Includes bibliographical references and index.
    ISBN 0–7914–4555–0 (hc: alk. paper).—ISBN 0–7914–4556–9
(pb: alk. paper)
    1. Technical writing—United States—History—20th century.
I. Title.   II. Series
T11.L65   2000
808'. 0666—dc21   99-38253
                                                            CIP

10 9 8 7 6 5 4 3 2 1

Contents

Acknowledgments

This study began in 1995 with a nagging irritation that virtually all technical writing textbooks began by apologizing for our discipline. In that incipient stage of research, discussions with Alan Nadel gave form to what became this history. He has been my critical conscience in this project.

When this history began to take form, Jim Zappen lent his historian's sensibility to it, helping me to refine my ideas and observations. Beth Britt proffered unlimited hours of theoretical discussion and camaraderie. In the later stages of manuscript development, Art Young contributed his sage advice to help this book project come to fruition. Reviewers Marilyn Cooper and Steven Katz gave me support, encouragement, and astute critiques of the manuscript in process. Will Worzel contributed his computer design talents for the book cover. Special thanks are extended to the editors at SUNY Press and to Jeffrey Waite, proprietor of DogEars Antiquarian Books in Hoosick Falls, New York.

A number of colleagues have given me their support, encouragement, references, and advice to help develop this history: Chris Boese, Antonio DiRenzo, Skip Eisiminger, Elaine Graham, Lillian Horlyck, Kevin Hunt, Merle Metcalfe, Sue Mings, Lee Odell, Barton Palmer, Elizabeth Shea, Charles Sides, Langdon Winner, Elizabethada Wright, Yingfan Zhang, and Muriel Zimmerman. Mike Vatalaro steadfastly offered personal support, although he would rather work in clay than words.

Finally, this book is dedicated to textbook authors throughout the century who have taken on the daunting task of putting technical writing knowledge into print.

Introduction

Transforming Language Into Science

But in a strict sense technology is discourse on technique. It involves the study of a technique, a philosophy or sociology of technique, instruction in technique.
　　　　　　　　—JACQUES ELLUL, *The Technological Bluff*, 1990

The manner in which human sense perception is organized, the medium in which it is accomplished, is determined not only by nature but by historical circumstances as well.
　　　　　　　　—WALTER BENJAMIN
"The Work of Art in the Age of Mechanical Reproduction," 1936

role of tech writing

Good technical writing is so clear that it is invisible. Yet technical writing is the mechanism that controls systems of management and discipline, thereby organizing the operations of modern institutions and the people within them. The invisibility of technical writing attests to its efficiency as a control mechanism because it works to shape our actions without displaying its methods for ready analysis.

When technical writing is made visible for study, it is often characterized as writing whose subject matter is technical, applied, practical, or functional. Taken in these simplest terms, technical writing conveys technical information. In its most accomplished form (according to this understanding of the term), technical writing functions as a neutral conduit for information generated by applied scientists. Audiences for this technical writing can either be familiar or unfamiliar with the subject matter—experts or non-experts in a particular field or discipline. In either instance, the perfect technical writing will not contaminate the pure meaning of the applied scientific knowledge.

Yet many technical writing experts would argue that technical writing has a more active social role than being merely a neutral conduit for information—technical writing mediates the transfer of technical information from an applied scientist to an end user of the technology. In this model, good technical writing will enable a user to effortlessly adopt and implement a technology for a practical purpose. Again, the perfect technical writing will be so transparent that a user will be able to understand its information without any miscommunications. In this understanding, technical writing is a simple collaborative effort in which writers mediate technology for users.

In its mediation of technology transfer from applied scientist to end user, however, technical writing serves a more active function than merely a neutral collaborator/conduit. Technical writing controls how technical knowledge is made. It allows for control of technical knowledge and its power within a society. It works to bring native or uneducated practices into the realm of theory and science. It is a mundane discourse practice working to enable some types of knowledge and practice, while disabling other possible knowledges and practices. Technical writing works to (de)stabilize knowledge and practice within institutional and societal systems. It is shaped by these systems, while simultaneously shaping them.

Spurious Coin traces the development of technical writing in the 20th-century United States. Its method of inquiry has heeded Thomas Kuhn's call for historians of science to "display the historical integrity of that science in its own time" (3). In this vein, I have attempted to reconstruct a cultural context for past technical writing practices in order to understand these past practices not as ill-fitting or quaint compared to contemporary understanding, but as legitimate practices within their situated contexts. Since technical writing deals in knowledge made through pure and applied science, the practice of communicating this knowledge can be seen as a scientific mechanism or apparatus for determining proper valuation and credit for the scientific product (knowledge).[1] As M. J. Mulkay observed, the rewards accruing to a scientist are "granted only to the scientist who first communicates the particular finding to the research community" (100). In their remarks introducing Price's article "The Relations Between Science and Technology," editors Gabor Strasser and Eugene Simons also found that scientific knowledge is made through communication: "Oversimplified, technology is what one is paid to do; science is what one acquires title to by giving it away, that is, publishing" (149). Karin Knorr Cetina theorized this knowledge-making as a "struggle for scientific capital" or "a struggle for the accumulation of 'say' or of 'holdings' in the scriptures as indicated by the number of papers, citations received, and ultimately, textbook entries" ("Producing" 687). In Knorr Cetina's model of scientific knowledge-making, "scriptures" are "a body of writings considered to be authoritative, and structured by indicators of differential value (the names of authors, journals, and publishing companies" ("Producing" 670). By communicating their knowledge, scientists seek to modify the scriptures and, thereby, the concepts that regulate further knowledge production. If a scientist's communication can modify these concepts, that person's knowledge is accorded value. This ability to transform knowledge into value is central to technical writing's social function.

WHAT IS SCIENCE? WHAT IS TECHNOLOGY?

As mentioned above, the term "technical writing" is most often defined as somehow dealing with technical information. In this history, however, I assume that technical writing deals in pure and applied scientific knowledge, as well as technical information, for both circulate through communication in an economy of scientific knowledge and power. As Andrew Pickering observed, "[K]nowledge is for use, not simply for contemplation, and actors have their own interests that instruments can serve well or ill" (4). Through the instrument of technical writing, scientists use knowledge to participate in an economy, gaining rewards and credit for their ideas if they are socially acceptable.[2] These rewards can be in the form of increased or continued funding, awards, publications, and elevated social status, to name a few.

This argument for including scientific knowledge as a subject of technical writing does not necessarily assume that scientific knowledge is substantively different from technical information. This distinction is more of a social convenience than an objective fact. In one sense, you could argue that applied scientific knowledge is more like technological information than it is like pure scientific knowledge. You could also argue, as M. J. Mulkay has, that distinctions between pure and applied science (or technology) are not always clear: "[T]he same research project may be defined quite differently by different social groupings. For instance, research which those scientists actively involved see as 'pure' will often be regarded as 'applied' the the officials providing the funds" (94). So once again, social and economic factors can (re)define what is pure and what is applied science based on economies of power and knowledge. Distinctions between pure and applied scientific knowledge—or science and technology—are not hard and fast.

The role of technology, like that of applied science, is to apply scientific knowledge to products and processes. E. Layton defined this process of technological development as having three stages: invention, innovation, and diffusion (198). This development is often seen as taking pure scientific knowledge and making it functional through these stages, as in Richard Barke's example of the development of magnetic resonance imaging from pure research on the study of magnetism associated with atomic nuclei (7). Yet this relationship between pure science and technology is a two-way street: functional developments can suggest areas for pure research (Price 153). So, once again, distinctions between pure science, applied science, and technology blur when their social and economic functions are examined.

Public and private funding agencies tend to spend money on applied, rather than pure, science, as noted by Barke (7) and Mulkay (94). If applied science is more likely to be funded than pure science, then in a social sense, pure science needs to be connected to applied science and technology to secure funding. And yet, by severing connections to technology, pure science can avoid social responsibilities for the products of their knowledge. As Derek de Solla Price claimed, distinctions between science and technology are most useful for gaining rewards or evading responsibility (150). Bruno Latour extended this argument, claiming that distinctions between the terms "science" and "technology" are "a figment of our imagination, or, more properly speaking, the outcome of attributing the whole responsibility for producing facts to a happy few" (174). Finding that distinctions between the terms "science" and "technology" serve socially "to settle responsibilities, to exclude the work of outsiders, and to keep a few leaders" (174), Latour argued that the term "technoscience" more accurately captures the role of science and technology in modern Western culture. While I will continue to use the terms "science" and "technology" throughout this history, I agree with Latour that distinctions between the two are not sharp and that in many ways the two terms can be used interchangeably when considering the social and economic functions of scientific knowledge and technical information.

MINTING THE COIN OF SCIENTIFIC KNOWLEDGE

Just as the most elevated science is pure, so, too, is the most accomplished technical writing purified into neutral invisibility. It is of these purified words that the clearest scientific knowledge is made. This purified scientific knowledge can then participate in an economy of knowledge and power—an economy unsullied by base words or ideas. Scientific knowledge becomes coinage within this economy and technical writing mints this coin. Thus the importance of purified technical writing: to keep the coinage free of base contaminations. Good technical writing ensures a genuinely sound currency for the scientific economy.

These notions of scientific knowledge, technical writing, and economic security run deep under current technical writing practices, forming a foundational metaphor for assigning value to knowledge through technical writing. This metaphor of technical writing as coinage is not new; it was first set in text by mining engineer-turned-editor T. A. Rickard at the turn of the 20th century. In a 1901 paper, Rickard foregrounded a discussion of technical language as the coinage of intellectual and scientific exchange, saying that the "growth of knowledge has required an increase in the medium of intellectual exchange" (111) or currency made available through the minting process of technical writing. Yet the nation's language was not pure enough to mint a genuine currency: "The nation debases its language with slang, with hybrid and foreign words, the impure alloys and cheap imports of its verbal coinage, mere tokens that should not be legal tender on the intellectual exchanges" (111–12). In Rickard's view, scientific and technical language should be purified of Latinate expressions, words imported from places other than England and the eastern United States, and literary meanings. Rickard set out an aesthetic of efficiency and purity for technical language, similar to what would be considered beautiful or valuable in the mechanics and products of a mining operation.

In Rickard's 1908 technical writing textbook,[3] the chapter "Spurious Coin" covered the debasing of technical language in mining and metallurgy through the adoption of foreign and "vulgar" terms. He argued that by debasing the coin of technical language, engineers threatened science, the survival of the human species, and an economy carried out through the currency of science and technical writing. The stamp of science gave technical language its economic value, just as the stamp of the King gave coinage its representational value as currency.

The fundamental ideas about technical writing that Rickard set out in his 1908 textbook stand today as common sense knowledge about technical writing practice. By setting this common sense against equally possible but (to us) uncommon notions of reality from other places and times, this cultural history will explore why technical writing works to (de)stabilize our common sense based on legitimated scientific knowledge. One of these bits of common sense holds that technical writing is an invisible conduit conveying applied scientific knowledge from developers to users. This seemingly obvious notion, though, contains conflicts for power through language—contextualized conflicts between contradictory notions of reality and their authority to legitimated knowledge.

This history traces the development of the system of knowledge and power that technical writing controls in the late-20th-century United States. It investigates problems inherent in the tension between scientific knowledge-making and liberal arts knowledge-making that render technical writing both the genuine and counterfeit coin of dominant scientific knowledge within our culture. This cultural history is constructed around a framework of five intellectual trends running through centuries of Western civilization: the use of clear, correct English; maximum efficiency of production and operation; the need to contribute to a general fund of scientific knowledge for the betterment of the human condition; tension between the role of science and art within a culture; and a redemptive urge to purify language and standardize practice. These intellectual trends sometimes appear clearly on the surface of discussions about science, technology, management, and communication. They sometimes go underground as implicit assumptions underpinning cultural debates. When these ideas lie on the surface of discourse, they can be easily gathered as resources for constructing this history. When they lie underground, they must be divined and mined in a more laborious process that can follow seemingly unrelated ideas to the lode.

This contextualized history will work to illuminate additional facets of existing studies within the field of technical and professional communication and will help researchers to understand why we consider certain topics in this field to be important. Chapter One explores a selection of studies in technical and professional communication to illustrate how this cultural history can explore silent voices within traditional objects of inquiry. Chapter Two begins to construct a cultural history of technical writing and textbooks as cultural artifacts. The background of ideas set out in this chapter provides a foundation from which current technical writing and pedagogical practices can be understood in a historical perspective. This chapter also discusses the works of Francis Bacon and how they served to replace scholastic speculation with scientific experimentation as the dominant knowledge. Chapter Three follows the development of the textbook and standardized education in response to Bacon's project for a public science. It also traces Bacon's ideas through the works of Locke, Hume, and Huxley. Chapter Four begins to address the impact of T. A. Rickard's technical writing textbook, *A Guide to Technical Writing* (1908), on a century of technical writing practice. Chapter Five explores how mechanical engineers developed management systems for large, complex organizations in the United States during the late-19th and early 20th centuries. Through their designs for social systems, these engineers shaped the contemporary knowledge system that relies on the dominance of scientific knowledge and is controlled through technical communications. Chapter Six focuses on Fredrick Taylor's scientific management system and its transformation into a functional-military management system based on principles of efficiency. This development of a specialized management system controlled by technical communication is reflected in George Crouch and Robert Zetler's textbook *A Guide to Technical Writing* (1948). Chapter Seven traces tension between engineers as technically trained scientific knowledge-makers and technical writers as liberal arts trained scientific knowledge-makers—a tension that developed after World War II with the separation of technology development and technical writing. When technical writing became a specialized profession in the 1950s, technical writing textbooks such as Kenneth Houp and Thomas Pearsall's *Reporting Technical Information* (1968) began to resemble traditional English composition textbooks. This development reflected an increasing influence of classical liberal arts knowledge upon which traditional writing instruction is based. The increasing influence of liberal arts in technical writing was met, however, with a conservative engineering systems approach to technical writing in J. C. Mathes and Dwight Stevenson's *Designing Technical Reports: Writing for Audiences in Organizations* (1976). Mathes and Stevenson's revival of engineering-based technical writing sought to replace the stamp of science onto the coin of technical writing, thereby transforming a counterfeit currency of liberal arts knowledge back into the genuine currency of science. Chapter Eight speculates on ways to revalue the currency of science by expanding the types of knowledge that technical writing can include in practice.

Stanley Carpenter has argued that "language and technology are not sharply distinguishable" (164) and he has a point that a culture's technology is another symbolic language expressing the culture's values and achievements. Yet in a culture dominated by scientific knowledge, the distinction between science and not-science—or scientific knowledge and language—is important to maintaining a stable knowledge/power system. Technical writers—whether they are specialist writers or scientists writing—negotiate this border between science and not-science as part of making scientific knowledge through language. Scientific knowledge can be made through unscientific language practices, but these practices must be transformed into science before they can confer the stamp of genuine currency onto the knowledge they mint. The tensions inherent in this transformation are the subject of this history of technical writing.

Spurious Coin

1

How Credit for Scientific Knowledge Is Appraised

Although transcending your origins in order to evaluate them has been the opening move in cultural criticism at least since Jeremiah, it is surely a mistake to take this move at face value; not so much because you can't really transcend your culture but because, if you could, you wouldn't have any terms of evaluation left—except, perhaps, theological ones.

—WILLIAM BENN MICHAELS,
The Gold Standard and the Logic of Naturalism, 1987

During what Samuel Florman has called "the Golden Age of Engineering" from 1850 to 1950 (6), technical writing gained authority as a discourse by virtue of its position as the lingua franca of engineering and scientific knowledge. Technical writing made engineering knowledge material, engineers were intellectual leaders in the United States, and technical writing was the currency representing the cultural exchange value of their practices. It is little wonder that, in this golden age of engineering, technical writing should begin to exhibit traits of a budding profession in the United States. Technical writing practices, for example, were compiled into textbooks. This act of compiling gave authority to the discourse of technical writing and worked to transform the status of engineering writing from a "sort of literary effeminacy" (Rickard, *Guide* 9; see also 129) to "one of the most valuable subjects you will study in college" (Anderson 4).

At the same time as this shift to professionalism occurred within technical writing, the role of technical writing within our culture remained largely unchanged. It continued throughout this century to stabilize our culture's system of knowledge and power based on scientific knowledge. Technical writing worked to control behaviors of people making knowledge in scientific laboratories, people making technological goods in factories, and people managing processes or services in factories or retail establishments. It was through technical texts that scientists received credit for their ideas and rewards that might accrue from these credits. Technical records tracked employee and machine production in factories, as well as distribution and sales of goods, thereby enabling managers to determine reward structures for employees and stockholders. Technical writing has been used to track the activities of people and machines, with the goal of assigning value to those activities. Technical writing is the control mechanism of scientific and technical knowledge production. As technical writing gained professionalism and power throughout the 20th century, it has become the subject of disciplinary study. Researchers in technical and professional communication have studied technical writing products and practices in order to improve the efficiency of this control mechanism. In the first half of the century, this study was primarily limited to grammatical advice, exposition of forms, and basic rhetorical principles of audience awareness. Most technical writing study was published in textbooks or handbooks aimed at an audience of engineers, managers, or engineering students. These were the people who practiced technical writing as part of their professions or professional preparation. Specialist technical writers did not appear until after World War II, when some organizations split some communication functions from research and development functions in order to make technology development more efficient. During this last half of the 20th century, technical writing developed as a specialized profession apart from the science and engineering professions. Although the subject matter of technical writing was scientific and technological knowledge, its practitioners were not necessarily scientists or engineers. Yet technical writing continued to control knowledge production in these fields and the rewards that accrued from this production.

WHO GETS CREDIT?

In a system of scientific knowledge production where individuals are rewarded for their ideas, technical writing is the apparatus for assigning credit and value for these ideas. In one aspect, technical writing separates out people who are qualified to accrue rewards for scientific knowledge from those who are not qualified. In this sense, technical writing refines the pool of people eligible for rewards accruing from the production of scientific knowledge. This smelting function is socially important to maintaining a system in which people with scientific and technical knowledge hold power to shape social decisions. As Jacques Ellul argued, in a society where people with technical or scientific knowledge can "direct the nation according to their technical competence" (24), specialized language becomes an important instrument for determining who has the technical competence to become eligible for that society's rewards. Only people who know the specialized language and can turn this knowledge into specialized practices are eligible for the power, influence, and funding that accrue from that knowledge. "This is one of the important aspects of the power that ordinary people do not share" (Ellul 27). Technical writing serves to stabilize this social distinction between people who have technical knowledge and those who do not. Technical writing is a tool for appraising people based on their knowledge, thereby working to ensure social stability.

This transactive nature of technical writing is not lost on practicing scientists, as evidenced by Bruno Latour and Steve Woolgar's interviews. They quoted one younger scientist estimating the value of his work in a knowledge economy: "'This instrument can bring me ten papers a year'" (190). Another scientist estimated the return on his research investment as communicated in writing: "'[M]y ability to find a job in research again will be increased in one year when the papers we are writing now will be published'" (191). These are two of many statements Latour and Woolgar heard in which scientists employed economic metaphors to describe their knowledge production. This weight of evidence suggests that many scientists envision themselves participating in an economy of scientific knowledge, where technical writing is the instrument for making their knowledge material for valuation by their peers. Through this writing, scientists work together to assign value and rewards to ideas that conform to the group's "forceful and coherent characterisations of their social and intellectual world" (Gilbert and Mulkay 137). In this world stabilized through the instrument of technical writing, scientists, in turn, become "malleable" agents working in an economy of scientific knowledge: "In the laboratory, scientists are 'methods' of going about inquiry; they are part of a field's research strategy and a technical device in the production of knowledge" (Knorr Cetina, "Couch" 119). When scientific workers can be shaped by the social power of technical writing, they become, like the writing itself, instruments of knowledge production and appraisal in a stabilized economic system.

WHAT RESEARCH DOES NOT SEE

Although technical writing is an important instrument for stabilizing an economy of scientific knowledge, this cultural role has not been adequately studied in technical writing research. After years of research, technical writing professionals cannot fully answer questions about how technical discourse participates in culturally grounded contests for knowledge and power. We cannot explain why ideas and practices that were legitimate less than 100 years ago are no longer legitimate. We do not understand how technical writing provides a currency for scientific knowledge. How can communication researchers uncover institutional systems of discourse formation that will help us address these uncovered issues? We can begin by examining how a research model based on critical theory provides a vocabulary and framework for researchers to discuss issues of knowledge and power. Two articles written by Lucille McCarthy, both dealing with the *Diagnostic and Statistical Manual* (*DSM*) charter document in psychiatry, provide a comparison through which we can see how research based on critical theory enables researchers to illuminate institutional relationships of knowledge and power reflected in discourse practices.

In the first of McCarthy's two articles, she examined how the constraints of the *DSM-III* shaped psychiatric practice. Using a social constructionist research framework, McCarthy's 1991 study resulted in a descriptive account of the psychiatric practices she observed in a hospital with the help of her friend, child psychiatrist Dr. Joan Page Gerring. In setting up the study situation near the beginning of her report, McCarthy alluded to silenced points of view in her explanation of how a biomedical model of psychiatry gained dominance over an interpretive model through a revision of the charter document: "*DSM-III* is a charter document is [sic] psychiatry, and the particular reality that it stabilizes is the biomedical conceptual model of mental illness" (359). McCarthy went on to explain how, in the biomedical model of psychiatry, "each patient exhibits a form of human activity which can be correlated with biological, psychological, and sociological variables" (362). She contrasted that to the interpretive model, in which "each patient presents 'an exercise in hermeneutics: a reading of the books of consciousness and behavior for their hidden meanings'" (362). What is unsaid in McCarthy's comparison is that the biomedical model of psychiatry allows for more quantified diagnoses, since patients' behaviors relate to scientifically described variables. Once a psychiatrist completes the *DSM-III* checklist describing the patient's behavior, that psychiatrist will be able to prescribe medications and other therapies, and determine how long the patient should be institutionalized or treated as an outpatient. This type of standardized diagnosis is far more amenable to insurance reimbursement, hospital administration, computerized record keeping, pharmaceutical monitoring, and psychiatrist treatment planning than psychiatric diagnoses rendered using the interpretive model. In addition, the biomedical model works to give psychiatrists the same professional status as medical doctors, who also use a biomedical model of physical illnesses for their diagnoses.

In her article on *DSM-III*, McCarthy described how the psychiatric profession ensured that a scientific biomedical model was valued over an interpretive model of psychoanalysis. But in using a case study approach and a social constructionist research framework, she was not able explore the political, economic, or social implications of her findings. Instead, McCarthy could simply describe the process of valuation without analyzing its cultural implications: "[I]t is certain that the dominant perspective of virtually all of the 130 members of the American Psychiatric Association task force which developed *DSM-III* was biomedical. These people were chosen on the basis of their clinical and research experience, and most had made 'significant contributions' to the literature in diagnosis" (362). The people who determined what counted as "'significant contributions' to the literature in diagnosis" were influenced by cultural pressures and conflicts that shaped the makeup of the *DSM-III* task force and, through that document, the practice of psychiatry—conflicts between practitioners of the biomedical and interpretive models of psychiatry, pressure to elevate the status of the psychiatric profession relative to the medical profession, pressure to receive a higher level of insurance funding, pressure to make record-keeping more efficient through the use of computers, conflicts between what is legitimated as scientific knowledge and what is marginalized as non-scientific lore.

The case study reported by McCarthy suggested enough cultural conflict to prompt a response from fellow communication researcher Carl Herndl. In his response, Herndl called for "new research to investigate the ideological work and the struggles that occur within professional discourse" ("Teaching Discourse" 361). This call for a study of technical writing practice within a cultural framework was based on the current state of social constructionist research in technical and professional communication, which is predominantly descriptive but not critical. By looking more closely and critically at tensions like those between the biomedical and interpretive models of psychiatry in McCarthy's article, technical communication researchers can uncover and analyze political and ideological contests in practices that work to legitimate some knowledge and marginalize others.

LOOKING AT TECHNICAL WRITING
THROUGH A CULTURAL STUDY FRAME

McCarthy and Gerring responded to Herndl's critique in a subsequent article, in which they took a critical approach to examining the political effects of the next revision of the *Diagnostic and Statistical Manual, DSM-IV*. Instead of describing the effects of *DSM-IV* on one hospital's organizational culture and making generalizations about charter documents based on this description, McCarthy and Gerring's critical study analyzed relationships of the revised *DSM* to other cultural elements within a situated context. They found that this particular document had the following effects: "(a) to further solidify the dominance of the biomedical model of mental disorder within psychiatry, (b) to maintain the position of psychiatry as the high-status profession among competing disciplines within the mental health field, and (c) to achieve acceptance of psychiatry as a mature, research-based specialty within medicine" (149). In their critical study, McCarthy and Gerring could talk about how the *DSM-IV* influenced trends in the psychiatric profession, how it affected relationships between psychiatry and other professions, how it worked to legitimate biomedical knowledge as dominant in the medical professions. These issues of change, power, and knowledge were tacit in these authors' first case study of the *DSM-III*. They came to light explicitly when the authors used a critical research approach to the *DSM-IV*.

Applying this critical approach to analyzing technical writing illuminates issues of power and knowledge in technical writing practices, as exemplified in McCarthy and Gerring's second article. Although a few researchers in technical and professional communication have begun to explore how technical writing practice and pedagogy are implicated in cultural contexts in which they are practiced,[1] the bulk of research in this field relies on a social constructionist paradigm that isolates the object of inquiry for the purpose of analysis. This social constructionist research paradigm gained prominence within composition studies in the 1980s in part as a compensation for earlier cognitivist studies—exemplified by the work of Flower and Hayes—that focused on universal psychological actions within "the mind" as a way to explain how individuals composed texts. The social constructionist approach allowed researchers to take into account influences on a writer that they saw as arising outside that writer's own mind, influences such as group or "community" affiliations,[2] organizational or professional affiliations,[3] or relationships between writers and readers.[4]

Some social constructionist researchers sought to synthesize their social position with the cognitivist position to come up with an approach that saw writing as stemming from an individual's mental representations of a social communication situation.[5] Other social constructionist researchers sought to combine their social position with one influential aspect of cultural contexts of communication—gender—to more fully explain nondominant patterns of communicating.[6] But even this inclusion of gender as one aspect of a cultural context for communication could not illuminate a wide range of cultural influences affecting a given communication. The social constructionist approach enabled researchers to talk about some influences external to an individual's mind that affect how that individual composes, such as group norms or specialized languages. But even with this recognition of social influences, the constructionist approach has two main limitations: (1) it maintains a decontextualized object of study[7] and (2) it does not allow researchers to address changes, contradictions, or conflicts in the object of study that can point to sites of ideological tension being played out in practice. It does not allow researchers to explore how technical writing practices work to legitimate some types of knowledge while marginalizing other possible knowledges.

An example of how the social constructionist research framework has been applied can be seen in an article by James Paradis, David Dobrin, and Richard Miller (1985) entitled "Writing at Exxon ITD: Notes on the Writing Environment of an R&D Organization." In this article, the authors looked at "what motivates research and development (R&D) employees to write and edit their internal work documents or how the industrial environment influences the way in which employees carry out these processes" (281). They computed totals and percentages of time spent writing by staff, supervisors, and managers; they looked at how documents are cycled among staff, supervisors, and managers; they looked at conflicts between staff/writers and supervisor/editors over close editing revisions. They found that writing is linked to job responsibilities, productivity, and notions of information transfer. They identified social functions in which writing participates, including work management, self-promotion, networking with a community of colleagues, accountability, idea stimulation, and self-education. But they kept these activities confined within the boundaries of Exxon, which was an organization isolated from its cultural context for the purpose of analysis. The notion of corporate culture used in this study was useful for describing existing practices as R&D staff wrote documents. But it did not allow an examination of the power structure underpinning existing practices in which supervisors' ideas must be followed by staff members or in which writing is seen as participating in larger issues of self-promotion and accountability (within and external to Exxon). This approach to the "social" missed how technical writing within Exxon worked to stabilize Exxon's position in relation to its competitors, the government, its customers, its suppliers, etc. This study could not analyze how Exxon's writing participated in an economy of technical knowledge and power.

The communication practices these researchers observed within Exxon were described, but the assumptions they were based on were misrecognized[8] as "natural" or inevitable practices. The R&D people at Exxon could not be seen as participating in a culture that extended outside the corporation, in which government contract bidding, legal liabilities, inter-corporation competition for profits, and public image played a part in their decisions about report writing—a culture where corporate practices are shaped by tensions among institutions. Although Paradis, Dobrin, and Miller gave us one view of how organizational communication is structured within Exxon, it is important to note that even a corporate culture extends beyond the corporation's walls. A wider cultural context exists prior to the R&D writers' observed practices, shapes their practices, and in turn is shaped by their practices. The social constructionist paradigm illustrated in this Exxon study does not provide a framework to set R&D practices at Exxon within a context of political, economic, ideological tensions.

In this and other articles, the authors misrecognized description for prescription and called for technical writing pedagogy that teaches students to reproduce the types of texts found in the organizations studied.[9] These prescriptions based on observed practices can help students make the transition from the academy to the corporation by preparing them to maintain existing knowledge, which is certainly an important component of workplace performance. But because these prescriptions do not adequately prepare students to deal with issues of difference and change within organizations, some cultural researchers have questioned the effectiveness of this approach.[10]

In one study that analyzed culture in an inclusive sense, Steven Katz illustrated the importance of difference and change by exploring discourse practices within the community of Nazi Germany. Katz argued that Hitler's rhetoric had its own ethic—the ethic of expediency—that formed a coherent foundation for his programs and propaganda. Katz analyzed texts written by Nazis to show how

[i]n Nazi Germany (and I will suggest, in our own culture) science and technology become the basis of a powerful ethical argument for

carrying out any program. Science and technology embody the ethos of objective detachment and truth, of power and capability, and thus the logical and ethical necessity . . . for their own existence and use ("Ethic of Expediency" 264).

In a traditional application of social constructionism, Katz would have described how Hitler constructed an an ethic of expedience that became standard discourse practice within his organizational community. He would have made generalizations about this discourse practice and would probably have recommended remedies to counteract its persuasive power through propaganda. But Katz chose to place Hitler's discourse within a historical context to understand how his propaganda valued technical knowledge while devaluing humanistic knowledge. Katz put forward this example to illuminate a similar condition in our own culture and serve as a warning against runaway technocracy. He used this extreme example to illustrate how an economy built single-mindedly on scientific knowledge can ultimately threaten our social systems. While admitting that the Nazi example is extreme, Katz clearly placed current technical writing practices in contests for power and knowledge legitimation, a research outcome that relies on a critical approach to the object of inquiry and could not be accomplished with conservative description alone.

That social constructionist studies of technical writing ignore political struggles does not place technical writing outside these struggles for knowledge legitimation. Instead, studies that ignore legitimation struggles tacitly conserve existing relations that technical writing stabilizes within an economy of scientific knowledge. In other words, when researchers do not explore tensions and struggles embodied within technical writing, their research may help to change surface features of the knowledge system, but it will not change power relationships within that system. Technical writing practices based on this conservative research will continue to stabilize an existing economy of scientific knowledge that technical writing controls. In this vein, Carl Herndl argued that pedagogy based on social constructionist notions of consensus-based community discourse conventions works to reproduce systems of power and knowledge, not to critique or change them: "[T]he idea of cultural (re)production and the theory of resistance present the major political challenge to the work of social or epistemic rhetoric in professional writing research" ("Teaching" 353). In calling for researchers to critically analyze discourse practices in addition to describing them, Herndl urged researchers to combine institutional critique with constructionist approaches to "see how discourse and the reality it constructs are shaped by the political, economic, and material interests of professions and the institutions they create" ("Teaching" 354). By re-viewing social constructionist research, like McCarthy's *DSM-III* case study, to look for sites of institutional contests for knowledge legitimation, researchers can locate places for critical analysis of how discourse practices participate in the politics of professional and institutional relationships.

An example of how we could push the boundaries of a social constructionist research framework to include cultural critique can be illustrated by examining the article "Social Context and Socially Constructed Texts: The Initiation of a Graduate Student into a Writing Research Community" by Carol Berkenkotter, Thomas N. Huckin, and John Ackerman. In this article, the authors described how a new Ph.D. student "Nate" at Carnegie Mellon University (CMU) changed his writing practices during his first three semesters in that university's rhetoric and composition program. Describing his context as "entering a new discourse community," the authors showed how Nate learned to use conventional textual forms in introductions to his research papers. They explained that Nate was a novice in this discourse community at the beginning of his first semester, even though he had been an English teacher prior to entering the CMU program. They found he struggled in "making the transition from *composition teacher* to *composition researcher* (i.e., from practitioner to specialist)" and that this struggle "involves a difficult passage from one academic culture to another" (Berkenkotter's italics, 211).

In this study, the view of CMU culture in which Nate learned how to make the transition from teacher to researcher was limited to the confines of one university (just as the Exxon study limited culture to the confines of one corporation). When the authors raised questions that pointed to the academy as an institution within a larger cultural context, they truncated the questions at the borders of the academy:

> How . . . do the sociopolitical constraints that govern the "manufacture of knowledge" (Knorr-Cetina) in this emerging field affect a graduate student's choice of research program? To what extent are the issues that concern composition teachers subsumed by the agendas of mentors as they join powerful research or scholarly enterprises, such as the one that we studied? How will the increasing graduate specialization in rhetorical studies and educational research affect the development of the canon within composition studies? (212).

As in the Exxon study, this research report left unexamined the roles of the academy, the disciplines of composition studies and technical writing, and practices of writing researchers as cultural agents participating in historically localized relations of power and knowledge. This study isolated observed practice and misrecognized it as "natural" and inevitable—as common sense.

Both the Exxon and the CMU articles assumed that discourse studies should be valued more highly within the humanities. Both assumed that adopting scientific approaches to their objects of inquiry would help increase the value of their findings in the eyes of their writing studies colleagues, in the eyes of their colleagues in the humanities, and hopefully in the eyes of their colleagues in the sciences and the institutional administrators who make funding decisions, as well as the corporations, non-profit foundations, and governmental agencies who fund university research. Both studies decontextualized observed practices and assumed that the purpose of teaching technical and professional writing in the academy is to prepare students to fit into existing practices in other institutions.

Because of their research design, these studies could not question how, at the particular moment of the study, the academy participated in cultural relationships with other institutions such as the government, business, and industry. Examples of these relationships could be found in research grants from government, non-profit organizations, and industry to support projects at the university; consulting positions in business and government filled by professors from the university; government and private sector scholarships funding student tuition and expenses at the university; publishing industry practices affecting the careers of professors whose promotions are based in part on their publishing records. These studies could not examine how pressures to win a government contract, for example, might have affected the writing practices at Exxon or how pressures to publish research reports may have affected the type of research valued at Carnegie Mellon University. Nor can these studies examine how people in the academy use difference and resistance to accomplish non-dominant, devalued, or non-legitimated practices within a historically situated institutional hegemony. We cannot see if Nate used his knowledge from teaching English to help him practice the discourse conventions of research at Carnegie Mellon. We cannot see if the writers at Exxon used sections from previously written documents to make their writing tasks less time-consuming or whether they asked colleagues at other corporations or in government for insights into writing situations. We cannot know whether the writing assignments were distributed equally among the staff, or whether some people had the power to reassign their writing tasks to other employees.

PUTTING TECHNICAL WRITING
PRACTICES IN CULTURAL CONTEXTS

Researchers in technical communication have only begun to explore how technical writing is involved within historically situated institutional relationships of knowledge and power—how some types of knowledge are valued and legitimated through technical writing practice, while other possible knowledges are devalued or excluded as marginal. Yet, as Vincent Leitch argued, institutions are active agents in creating and (de)stabilizing systems of knowledge and power: "Institutions include . . . both material forms and mechanisms of production, distribution and consumption and ideological norms and protocols shaping the reception, comprehension, and application of discourse" (127–28). If the purpose of critical research is to understand and improve the practice under inquiry—or to facilitate change in a knowledge system—institutions where technical writing is practiced need to be reconstructed as cultural agents that are not necessarily bounded by any one organization's walls.

A limited view of culture does not allow writers of conservative research reports or histories to question assumptions about technical writing practices. In Stephen Doheny-Farina's exploration of technology transfer *Rhetoric, Innovation, Technology*, for example, technical writing was seen not as a transfer of information, but as participating in a "series of personal constructions and reconstructions of knowledge, expertise, and technologies by the participants attempting to adapt technological innovations for social uses" (ix). Although Doheny-Farina placed technical writing practice in a social setting, he assumed that problems in technology transfer have more to do with the quality of the technical texts than with economic, political, or social pressures or conflicts affecting a situated writing practice. In his analysis of McCarthy's case study of the *DSM-III*, Doheny-Farina generalized that "the charter document stabilizes the actions of the members of the discipline and the ways that they think about issues in the discipline" (26), thus forming a "constraint that gives shape to a discipline" (27). While it may be accurate to say that a charter document such as the *DSM-III* forms a constraint that shapes a discipline, this view of McCarthy's study focuses on the "how" of the document and does not ask "why." In other words, Doheny-Farina can describe how a text shapes practice, but does not question why the text includes the information that it does and not other information that would be equally possible to include—why the text legitimates some kinds of knowledge and not others. And further, what systems of power does the knowledge legitimated in the text uphold and what other possible systems of power does the text make impossible?

Similarly, when Doheny-Farina analyzed the Paradis, Dobrin, and Miller study of "Writing at Exxon ITD," he found that their writing practice "becomes part of the process of developing an organizational identity; it becomes part of the process of group membership" (*Rhetoric, Innovation* 28). Social influences that form the culture under investigation in this analysis of the Exxon study are limited to the interior of the Exxon Corporation. The "social" in "social constructionist" here extends to a group of people within one organization. The limitations of confining culture within the walls of one organization are illustrated in Doheny-Farina's consideration of a 1991 article by Herndl, Fennell, and Miller in which these authors examined miscommunication and misunderstanding in the Three Mile Island and *Challenger* disasters. Doheny-Farina constructed the communication problems that contributed to the *Challenger* disaster as problems of technology mediation: "Issues of texts miscommunication and misunderstanding . . . involve the failures of texts to mediate technology to users" (*Rhetoric, Innovation* 28). This focus on the technical writer's role as mediator between technicians and users allowed Doheny-Farina to explore how technical writers uphold science and technology's dominant knowledge economy within our culture. But it did not question what influences other than miscommunication might have contributed to the misunderstandings described in the Herndl et al. article. For example, the decision to go ahead with the *Challenger* launch was made despite evidence of previous O-ring erosion and a forecast of record cold weather at launch. Morton Thiokol engineer Roger Boisjoly argued for safety and against the launch at a meeting with NASA representatives the night before the *Challenger* was launched. He later suggested an extra-textual factor creating pressure to launch: Morton Thiokol was in the process of negotiating a $1 billion contract with the U. S. government for space shuttle parts and the government was considering second-sourcing Morton Thiokol. NASA staff were intent on launching and, during the pre-launch meeting, asked Morton Thiokol management to "rethink" their recommendation not to launch. Morton Thiokol reversed its no-launch recommendation to acquiesce to NASA's wishes, despite Morton Thiokol engineers' warnings about unsafe launch conditions. More than a simple case of misunderstanding, discourse surrounding the decision to launch the *Challenger* was influenced by economic and political considerations that legitimated some knowledge (the managers' judgment and decision to launch) and marginalized other knowledge (the engineers' data and warnings).

The knowledge legitimated in this case was clearly participating in an economy of institutional power.

If the "social" in a social constructionist framework generally resides within an affiliated group, "culture" within this research framework is similarly limited to reside within an autonomous group.[11] In an example of separating a governmental agency's discourse practices from their context of political influences, Susan Kleimann assumed that culture resides within a governmental organization (the General Accounting Office) in her study entitled "The Reciprocal Relationship of Workplace Culture and Review." Kleimann confined the idea of culture either to divisions within the General Accounting Office (GAO) or to the GAO as an isolated governmental entity and described how the organization's values influence document review practices:

> Two major influences shape aspects of GAO's culture. First, GAO reports often result in changes to national policy, legislation, and funding; second, until recently, most GAO employees were educated as accountants, valuing minutiae and accuracy. Consequently, the agency has a cautious culture that demands maintaining detailed and extensive workpapers, referencing all facts to these workpapers, wanting both accuracy and objectivity and requiring an extensive review process (58).

While Kleimann noted that GAO texts are influential in shaping national policy, legislation, and funding decisions, she confined the culture discussed in her study within the boundaries of the GAO. While she obviously studied one of the major institutions in any culture (the government), she did not ask how national and international political tensions shaped GAO texts and how politics were in turn shaped by GAO texts. She studied governmental technical writing, but did not place it in an international economy of knowledge.

The limited and conservative view of culture found in most research in technical and professional communication was articulated in Jack Selzer's description of intertextuality: "Indeed, 'context' or 'environment' or 'setting' or 'culture' might be understood as nothing more than a complex of language and texts, and individuals within an environment therefore might be understood as minds assimilated into its concepts and terminology" (172). Or further: "Readers and writers experience the flow of culture as a kind of collaboration among seen and unseen authors and texts and readers, all in the process of making sense. One job of the critic of culture . . . is to uncover the various resonances inscribed in the tapestry of text and to account for their source, their intricacy, and their meaning" (179). While Selzer's view of culture did acknowledge that cultural criticism involves itself with meaning, his view did not involve texts in tensions within situated ethico-political relations of power and knowledge—where knowledge legitimation is contested by various interested groups and "making sense" means something different depending on your point of view. To view culture as a "kind of collaboration" works to sanitize what Walter Benjamin described as barbarism inherent in the spoils of war (for cultural legitimation):

> Whoever has emerged victorious participates to this day in the triumphal procession in which the present rulers step over those who are lying prostrate. According to traditional practice, the spoils are carried along in the procession. They are called cultural treasures . . . There is no document of civilization which is not at the same time a document of barbarism. And just as such a document is not free of barbarism, barbarism taints also the manner in which it was transmitted from one owner to another ("Theses" 256).

"Making sense" within a framework of contests for knowledge legitimation is not merely a "kind of collaboration." From a critical point of view, making sense for the victor is not making sense for the vanquished, who might ask why their knowledge must be silenced.

In *The Differend*, Jean-François Lyotard described the silencing of non-legitimated or devalued knowledge as a "wrong" suffered in "a case of conflict, between (at least) two parties, that cannot be equitably resolved for lack of a rule of judgment applicable to both arguments" (xi). Because there is no universal rule for equitable judgment, actions taken through discourse must privilege one way of knowing over other possible ways of knowing. Unlike a simple idea of consensus-based collaboration, Lyotard's theory of discourse production holds that power is distributed unevenly among possible ways of knowing. Basing his description on how phrases are linked in discourse, Lyotard found, "In the absence of a phrase regimen or of a genre of discourse that enjoys a universal authority to decide, does not the linkage (whichever one it is) necessarily wrong the regimens or genres whose possible phrases remain unactualized?" (xii). Using Lyotard's theory, discourse becomes a contest for legitimating knowledge and culture is more hegemonic than simply collaborative. Discourse becomes a struggle mediated by culture. Technical writing participates in that struggle by working to assign value to scientific knowledge, thereby minting the currency for its economy. Devalued knowledge, like a counterfeit coin, will not circulate widely in this economy; highly valued knowledge will circulate widely as the genuine coin.

Struggles for value are contained within technical writing. For Michel Foucault, discourse holds histories of struggles for knowledge legitimation and the articulated discourse subsumes other discourses that were possible, but not articulated. In arguing for the study of culture through discourse analysis, Foucault described how the legitimated knowledges articulated in discourse embody historical struggles for their legitimation and conquest:

> In the two cases—in the case of the erudite as in that of the disqualified knowledges—with what in fact were these buried, subjugated knowledges really concerned? They were concerned with a *historical knowledge of struggles*. In the specialized areas of erudition as in the disqualified, popular knowledge there lay the memory of hostile encounters which even up to this day have been confined to the margins of knowledge (Foucault's italics, *Power/Knowledge* 83).

At the margins of knowledge we can find two types of delegitimated knowledges: erudite learning that may have been previously legitimated knowledge, but has been subsumed (or conquered) by other subsequently legitimated knowledges; and naive "know-how"[12] that was previously legitimated as sufficient for carrying out everyday practices, but also has been subsumed by other subsequently legitimated knowledges (usually some sort of science or theory). Technical writing has worked to educate know-how into science through its own technologies of language. This educated know-how can then participate in an economy of scientific knowledge and a culture of technology.

Technical communication, rather than being seen as a simple collaborative effort in which writers mediate technology for users, can be seen as working to legitimate and value some kinds of knowledge while marginalizing and devaluing other possible knowledges. Because technical communication participates in institutional relationships, it works to organize knowledge through science and practice through theory. This organizing activity is found by Michel de Certeau to be a trend in Western culture since the time of Francis Bacon: "[T]he sciences are the operational languages whose grammar and syntax form constructed, regulated, and thus writeable, systems; the arts are techniques that await an enlightened knowledge they currently lack" (*Everyday Life* 66). Because science forms the legitimated language of practice, it is a "writing that conquers" (*Writing of History* xxv) other practices based on naive "know-how":

But at the same time that they acknowledge in these practices a kind of knowledge preceding that of the scientists, they have to release it from its "improper" language and invert into a "proper" discourse the erroneous expression of "marvels" that are already present in everyday ways of operating. Science will make princesses out of all these Cinderellas. The principle of an ethnological operation on practices is thus formulated: their *social isolation* calls for a sort of "education" which, through a *linguistic inversion*, introduces them into the field of scientific written language (de Certeau's italics, *Everyday Life* 67).

If technical communication is the mediator between technology and what we have come to term "users," technical communication practices work to conquer users' naive know-how and reformulate these naive practices into scientific discourse. In so doing, technical communication participates in a writing that conquers naive knowledge by educating it into the technologies of scientific disciplines. Thus, technical writing participates in an economy of scientific knowledge and power within our culture—an economy that can only be illuminated using critical approaches to discourse practices.

A few recent studies in technical and professional communication point to an approach for analyzing professional communication and composition from a historically situated perspective. Richard Freed advocated widening our views of discourse communities beyond "company specific" boundaries to enable analysis of inter-company relationships (213, n. 8), thereby bringing corporate relationships into play and expanding culture beyond one organization. Freed's work also illustrates how using Lyotard's notions of grand narratives and *petit recits* as set out in *The Postmodern Condition* enables researchers to reconceptualize knowledge as historically situated: "Because the shape and tonality of knowledge vary by locale, and because for that locale the tone and temper of its knowledge rings and feels true, truths at one locale may be different from those held self-evident at other sites and from those held at different times at the same locale" (204). Using critical theory allows Freed to explore how technical and professional communication work with situated knowledge that is profoundly shaped by contests for legitimation and which, in turn, shapes subsequent discourse and knowledge.

Other researchers have begun to consider how discourse practices participate in institutional relationships. For example, Bruce Herzberg advocated that composition researchers use Foucault's archaeological research approach "to analyze more closely the role of our institutions and disciplines in producing discourse, knowledge, and power" (80).

He asked this question of the social constructionist research paradigm: "[W]hen the group agrees on standards for sufficient evidence or adequate organization or coherent argument, what is the source of its authority?" (79). This questioning of the basis of authority opens up discussions of how discourse participates in power/knowledge systems. In other words, what power sanctions the authority of the knowledge that is described in observed discourse practices? Ben and Marthalee Barton also asked this question of power and authority in exploring the practice of cartography. They found that maps as discourse practices were closely linked to institutional systems of power and knowledge: "Ultimately, the map in particular and, by implication, visual representations in general are seen as complicit with social-control mechanisms inextricably linked to power and authority" (53). These examples of institutional critique illustrate how discourse can be seen as participating in economies of knowledge and power, exploring why some knowledge is articulated and legitimated while other possible knowledge is marginalized or left silent.

In another example of how critical theory can inform writing research, Lester Faigley explored how subject positions we constitute in composition classrooms rely on teachers' roles "as representatives of institutional authority" (*Fragments* 130). By applying Foucault's archaeology method, Faigley found a technology of confession in writing classrooms where personal narratives are seen as productive of "truths." This technology reproduces existing relations of knowledge and power between teachers and students, in which teachers dominate and students are dominated: "Such an assignment of authority through a teacher's claim to recognize truth is characteristic of Foucault's description of the modern exercise of power. Foucault writes that power is most effective when it is least visible" (*Fragments* 131). Using Foucault's archaeology enabled Faigley to discuss issues of institutional power and individual subject positions in the composition classroom because Foucault's work, and critical theory in general, provide a theoretical basis for recognizing institutional relationships and a vocabulary for discussing power/knowledge systems.

Recent cultural studies of technical communication and composition point to the fruitfulness of an approach based on Foucault's archaeological research methods and augmented by closely related lines of critical theory. Applied to technical communication, this approach can illuminate how struggles for knowledge legitimation taking place within technical writing practices are influenced by institutional, political, economic, and/or social relationships, pressures, and tensions within cultural contexts that transcend any one affiliated group. This

type of cultural study can help to answer questions about why technical writing practices work to value some types of knowledge while
devaluing other possible knowledges.

In order to look at struggles for knowledge legitimation that take
place within technical communication, researchers can begin by asking
Foucault's question, "How is it that one particular statement appeared
rather than another?" (*Archaeology* 27). The statements that did appear
in technical texts retell stories of the struggles, contradictions, and tensions within historic relations of knowledge and power. These statements also hold the silence of other statements that were possible but
did not appear in technical texts at the particular time and place under
study. By looking at statements that did appear and positing possible
statements that did not appear, the genealogical historian can construct
what Foucault called a "systematic history of discourses" (*Birth* 14).
Such a systematic history (or genealogy) of discourse asks questions
about how one possible discourse was produced and legitimated as
knowledge through technical writing, while other possible discourses
were not produced and legitimated.

ISSUES IN TECHNICAL WRITING
RAISED THROUGH CULTURAL STUDY

This question of how one group's discourse became knowledge within
a historically situated culture while another group's discourse was not
seen as knowledge strikes at the heart of current discussions of multiculturalism,[13] gender issues,[14] conflict,[15] ethics,[16] community,[17] and
postmodernism[18] within technical and professional communication. In
more traditional research designs, these issues are categorized in the
familiar language used above. When framed through cultural critique,
however, the language used to describe these issues will recast them in
poststructural terms. For example, issues of difference (as raised in all
the categories listed above) can be recast in these poststructural terms:
"Why has common sense about technical writing taken the form it has
when other forms of common sense were equally possible? Whose
knowledge gained power?" A related wording of this issue could ask,
"Why do technical writing practices work to reproduce our culture's
power/knowledge system? What ethical issues are at stake in this reproduction?" Issues of technical writing as a discourse could be
worded in these terms: "Does technical writing work as a double agent
in our cultural language wars between scientific and artistic knowledge legitimation?"

Looking at these questions in more depth, a researcher could question why specific aspects of technical writing practice were shaped as they were through institutional relationships with science, technology, and engineering. Looking at the role of theory in technical writing practice and knowledge, a cultural researcher could ask, "Why is theory implicit in technical writing? How might technical writing's cultural relations be changed if theory was explicit?" Looking at relationships between technical writing and critical analysis, a researcher could ask, "Why does technical writing restrain critical analysis in favor of clarity, efficiency, certainty, and authority?" Looking at relationships between technical writing and history, a researcher could ask, "Is technical writing (a)historical?" This history of technical writing's development will focus on these recast poststructural issues reflecting the categories researchers are currently exploring in the field of professional communication.

2

Technical Writing as the Lingua Franca in a Golden Age of Engineering

It may also be asked, in doubt rather than criticism, whether I am speaking of natural philosophy only, or whether I mean that the other sciences—logic, ethics, politics—should also be carried on by my method. I would answer that I certainly do think my words have a universal application. . . . For I am compiling a history and tables of discovery about anger, fear, shame and the like, and also about political matters, and no less about the mental actions of memory, composition and division, judgement and the rest; about all these, just as much as about hot and cold, or light, or vegetation or the like.
—FRANCIS BACON, *Novum Organum*, 1620

Because scientific knowledge dominates in 20th-century United States, technical language is the coin of the realm, circulating in an economy of scientific knowledge. And because technical writing's roots are most deeply planted in the field of mining engineering, with its emphasis on economics, value, and social stability, technical language conveys estimates of the value of ideas—estimates that originate from science as the supreme authority. In the tradition of Bacon's public science, technical writing participates in a social system that was established to democratize knowledge, taking experiential secrets for manipulating nature out of the realm of magic and making them legitimate subjects for scientific experimentation. Since this project for public science was vast, many workers were needed to accumulate a complete body of scientific knowledge about nature. Each scientific worker had the obligation to contribute his or her knowledge to this accumulation.

The links among technical writing, estimates of economic value, and public science surfaced in the first modern technical writing text written at the turn of the century by T. A. Rickard, who was a miner turned technical writer and editor. In elaborating his idea of a "general fund" of scientific knowledge, Rickard used economic metaphors to explain how technical writers participate in social systems:

> Yet each man possesses some little bit of knowledge, whether as observation, theory, or experience, that is his very own. Thus each can contribute something to the general fund; and seeing how much he owes, it is asking but little that he give cheerfully what he can. Of course, narrow minds still continue to fondle the mean belief that to give information gratuitously is to throw away a stock in trade, and that to keep secret the professional or technical experience of a life is to possess an added weapon in the arena of industry. But this is a pitiable fallacy scarcely worthy of castigation. If adopted universally we would be today as the Hottentot or the Eskimo; civilization has been evolved by the free exchange of thought and the frank transmission of experience. Whether we be advocates of free trade, fair trade, or reciprocity in matters of national industry, let us at least reject the shriveling policy of protection as applied to the worldwide traffic in ideas (130).

In the social system Rickard described, scientific and technological knowledge is the currency that keeps society's economy circulating. More particularly, scientific knowledge translated into technical language is a coinage that engineers should contribute cheerfully to the general fund of scientific knowledge for the betterment of living conditions.

Rickard's metaphor of scientific knowledge as circulating currency promised an increased store of knowledge through its free exchange. Foucault described this theory of exchange in monetary terms: "[W]hen coinage becomes more plentiful, as a result of a good circulation and a favourable balance, one can attract fresh merchandise and increase both agriculture and manufacturing. As Horneck puts it, gold and silver 'are the purest part of our blood, the marrow of our strength', 'the most indispensible instruments of human activity and of our existence'" (*Order* 178–79). Scientific and technical knowledge is wealth made visible through the coinage of technical language. By circulating the coinage of technical language, scientists and engineers can attract fresh ideas and "increase both agriculture and manufacturing" for the betterment of living conditions. Hoarding the coin of technical language, on the other hand, will shrivel agriculture and manufacturing, degrade living conditions, and even doom (scientifically advanced) humankind to low status on the evolutionary ladder, like the uncivilized "Hottentot or the Eskimo." Rickard found this hoarding, in the Hermetic tradition of the books of secrets, "coarse," "narrow," "mean," and "pitiable" because technical writing works to ensure the evolutionary progress of civilization—or social Darwinism.

Rickard's ideas about technical writing's role in Western cultures were articulated in what Robert Connors calls "the first technical writing textbook" ("The Rise" 332) originally published in 1908. To understand the context for this historical moment in technical writing, however, we must first consider how modern ideas of science and technology gained dominance over scholastic, speculative knowledge that was dominant in 16th-century England. This transformation of dominant knowledge, largely carried out by Francis Bacon and his followers, gave the term "technical writing" meaning through its relationship to scientific knowledge. This meaning was further articulated in John Locke's ideas of conveying scientific knowledge using clear, concise language, in granting colonial property rights to settlers who could gain the most conveniences from the land, and in protecting a stable English coinage through the use of colonial resources. Locke strengthened the ties between technical writing and economics during the 17th century, but it was Thomas Huxley's interpretation of Hume in the 19th century that solidified the separation of science from philosophy and religion, a separation also seen in Locke's work. By viewing empirical science as the only possible way to knowledge, Hume and Huxley held that certain knowledge was not based on *a priori* authority, but instead was gained only through first-hand observation and scientific proof. Huxley further differentiated between "pure" scientific knowledge, truths that scientists pursued from motives unsullied by practical considerations, and "applied" scientific knowledge, which partook in responsibilities for social outcomes. Through this distinction, Huxley shielded "pure" scientists from the social outcomes of their work and created a role for technical communication that was allied mainly to applied science and practical outcomes. It was this practical role for technical writing that Huxley's student, mining engineer T. A. Rickard, adopted in his technical writing textbook at the turn of the 20th century—well after empirical scientific knowledge had gained dominance over scholastic speculation and *a priori* textual authority as science.

WHAT KIND OF KNOWLEDGE GETS DEPOSITED IN TEXTBOOKS?

One seemingly inevitable aspect of technical writing practice, in the late-20th century, is that its professional practitioners work in a discipline taught within the academy as a coherent body of knowledge. Teachers of technical writing, therefore, have an array of textbooks from which to choose when planning their courses. These textbooks seem like an inevitable component of the academic environment. Teachers seldom question *why* textbooks are used. This simple question, however, points toward a crucial starting place for the history of technical writing written here, which will analyze technical writing practices based on textbook accounts of how technical writing should be carried out. To understand the kinds of knowledge deposited in textbooks as situated discourses, we must first realize that textbooks are cultural artifacts participating in knowledge/power systems.

Although compilations of knowledge were used as teaching aids for centuries, modern historians do not consider books before the 17th century to be "textbooks." For example, Pliny the Elder's *Natural History* (A.D. 77) is considered an example of Roman encyclopedism and not an example of an early textbook, even though this work was used in schools for centuries. Books of secrets, containing compilations of practical information about science and magic widely read by scholars from the early centuries A.D. through the Renaissance, are compilations or handbooks, but not textbooks. What is the defining characteristic of what we have come to know as "textbooks"? Why did "textbooks" become necessary at a certain point in history? What social function do "textbooks" achieve? To explore these questions is to begin to understand what kinds of knowledge gets deposited in textbooks and how that knowledge functions in a culture's economy of knowledge.

Textbooks first appear in histories of education in the 17th century with the work of Johann Amos Comenius, a Moravian educational reformer influenced by Francis Bacon. His books designed for students were part of a project to further Bacon's notions of public science for the improvement of the human condition. Following Bacon's call for a large cadré of workers who could carry out the work of public science, Comenius envisioned a centralized educational institution that included graded levels of education and disciplined knowledge. He devised textbooks as tools to discipline education for the purpose of disseminating useful knowledge and efficiently training future workers in an effort to establish a public practice of science. To better understand the implications of Comenius' textbooks in the 17th century, however, a historical review is needed of other concepts of books that, while not identified by historians as textbooks themselves, were forerunners of the textbook.

An early notion of compiling knowledge can be found in the encyclopedic practices of Hellenistic Greece and the early Roman empire. The Greek practice of encyclopedism stemmed from a move to a standardized educational curriculum in the Hellinistic Age. During this period, original scholarship was differentiated from popularized accounts of scholarship, and the latter were compiled into encyclopedias. These popular compilations attempted to cover all knowledge and were intended for use in a polysophistic curriculum: "In the Hellenistic Age the curriculum became more or less standardized and the concept of an *enkuklios paideia*, a circle of knowledge, became common" (Wagner 10). The term *enkuklios paideia* was later Latinized into the term encyclopedia. Thus, the encyclopedic tradition was closely linked to a project for standardizing a polysophistic curriculum.

The Greek encyclopedic tradition is illustrated by the works of the Stoic scholar Posidonius, who ran a fashionable school in Rhodes in the first century B.C. Describing Posidonius as a "popularizer of Hellenistic science," Wagner elaborated on Posidonius' work as an encyclopedist: "Posidonius belonged to the long-standing tradition of the polymath, or universal scholar. The term 'polymath' was generally used perjoratively . . . yet the tradition also included Aristotle, with whom Posidonius can be compared in influence" (14). Posidonius was a scholar who practiced original scientific research, but his scientific accomplishments were tarnished by his popular encyclopedic compilations and his reputation as a polymath. *[handwritten: —? One of encyclopedic learning]*

The era in which Posidonius lived saw Roman conquests and the decline of Greek civilization. It was also an era in which scholars were overwhelmed with the magnitude of extant texts. One way of coping with this accumulation of knowledge was to compile handbooks of practical recipes gleaned from original scholarship. These handbooks proved especially attractive to Roman scholars, who were generally impatient with theory and craved information that could be easily translated into action (Stahl 45). The political climate in the second century B.C. favored the typically Roman interest in action and correspondingly downplayed the typically Greek interest in otherworldly theorizing. Many Greek writers, such as Posidonius, adapted their interests to those of the dominant Roman culture and emphasized the encyclopedic tradition of compiling useful knowledge from extant sources, including recipes without the reasoning that originally led to them. Because these handbook compilations did not rely on reasoning in presenting their advice, many compilations juxtaposed contradictory or illogical information without needing to address or acknowledge theoretical problems. This emphasis on encyclopedic handbooks, "more than any other factor, was responsible for the subsequent decay of ancient science" (Stahl 30)

In the early centuries A.D., the handbook tradition was joined by another intellectual pseudo-tradition that was to shape the content of

handbooks—Hermetics. The original Hermetic texts, written in the
first to third centuries A.D., purported to be accounts of secrets of nature revealed in ancient times by the Egyptian god Thoth or Theuth. By
one account, "God had revealed a true knowledge of His creation and
its secret processes to the Jews. The 'ancient theology' had passed from
Moses to Hermes during their exile in Egypt. . . . it had been learned by
Pythagoras, who took it back to Greece, where Plato had been its last
exponent before its submergence under Aristotelianism" (Pumphrey
59). In another account, the original texts from Thoth were supposedly
revealed to the Chaldeans and ancient Egyptians, then to the Greeks
and Romans. This revelation was then passed on to the Arabs and to
the Europeans by the 12th century A.D. (Eamon 39). The original Hermetic texts were apocryphally attributed to ancient scholars and magi
to give the texts authority. This authority of divine revelation, purportedly passed through ancient magi, was especially powerful in the early
centuries A.D. when Greek rationalism waned and the promise of manipulating nature through revealed knowledge offered an attractive alternative. The content of these books—recipes for manipulating nature—was also attractive in a Roman culture that emphasized practical
action and results.[1]

A body of Greek Hermetic literature amassed that was "concerned
with astrology and the occult sciences, with the secret virtues of plants
and stones and the sympathetic magic based on knowledge of such
virtues, with the making of talismans for drawing down the powers of
the stars, and so on" (Yates 2). The Hermetic secrets contained in this
literature were intended for an initiated group of magi, who were to
pass on the secrets only to worthy students and were otherwise to protect them from the vulgar masses. These Hermetic writings paradoxically popularized the secrets of nature in two classes of literature: philosophical revelations and technical treatises on "astrology, alchemy,
natural history, medicine, and magic" (Eamon 18). Because the technical Hermetic literature promised practical domination of nature's secrets, it was more popular than the philosophical literature among
action-oriented Romans.

The popularity of Hermetic literature continued throughout Roman times and into medieval Europe. In its medieval manifestation,
Hermeticism distinguished between popular knowledge called science
and esoteric knowledge called magic, the latter of which was contained in the books of secrets: "In the Latin West the distinction
between the two kinds of knowledge was essentially that between
scientia and *magia*: science, the knowledge of the causes of natural phenomena, and magic, consisting of the techniques by which nature is

controlled, manipulated, and made to serve human ends" (Eamon 43). These books of secrets were enormously popular in medieval universities, where liberal arts students could supplement the formal curriculum with practical, "magical" information geared to making them employable bureaucrats in developing nation-states.

For example, a highly influential 13th-century book of secrets—the *Secretum secretorum* (*The Secret of Secrets*), purportedly letters of esoteric knowledge written by Aristotle to his student Alexander—dealt with statecraft as well as other standard Hermetic topics. Because the book promised to reveal esoteric knowledge of statecraft that would enable the knower to manipulate situations to the ruler's advantage, eager liberal arts students prepared themselves for successful careers by studying these clandestine texts:

> The distinction between public knowledge and esoteric council was, in reality, the secret of power and advancement for late-medieval intellectuals. As ecclesiastical and secular administration grew increasingly centralized, developing more complex bureaucratic structures, the need for trained manpower grew accordingly. But as the demand for a skilled work force increased, so too did the number of university graduates, thus intensifying the competition for positions. The *Secretum* entered the Latin West in the wake of political and economic changes that created a magnetic field for scholars, who, though generally of nonnoble status, were taking their places in a power structure that was traditionally reserved for the nobility. . . . Knowing the secrets of effective government was for them an essential tool for success (Eamon 50).

Books of secrets grew out of a Hermetic pseudo-tradition of revealed esoteric knowledge about manipulating nature—knowledge purportedly intended for an initiated few. The writing, publication, and distribution of the books belied this intention and instead participated in social systems that served, in part, to prepare university students for employment in centralizing medieval bureaucracies. While dealing with knowledge of statecraft, these books also addressed topics that could help nation-states and their rulers prosper, such as metallurgy, dye making, geography, and astrology.

In the Roman encyclopedic tradition, books of secrets were often indiscriminate in the information they included and were not especially concerned with exhibiting a coherent—or any—philosophy. They sought a pansophic scope, which became a problem by the 16th century when printing presses and global exploration resulted in masses of new information and artifacts. During this period, as in Greece during the first century B.C., the large number of artifacts and texts seemed overwhelming to Renaissance scholars, who sought to order knowledge through encyclopedic compilations. At the same time, a resurgence of Hermeticism during the Renaissance was evidenced in compilations that included revealed "secrets" from the Hermetic texts, as well as observed and/or experienced accounts in the rationalist tradition:

> Sixteenth-century scholars were not concerned to draw clear boundaries between the past and the present, reality and fantasy, the arts and sciences, because in their time these categories were neither firmly established, nor yet deemed the necessary prerequisite of meaning. . . . Everything is worth collecting, because everything may one day serve to make a show of learning, which is the same as being learned. In preparation for the vicissitudes of a public career, in order to acquire fortune and renown, the new scholars must amass a vast store of learning, to be copiously exhibited at the appropriate moment: what started off as the pursuit of knowledge for its own sake, ended up as a means of letting people know how knowledgeable they were. Whence the dual function of scholarly compilations: on the one hand the service rendered to the reader, on the other the celebration of the compiler's erudition and, further along the line, that of the reader who will in turn be able to display his learning (Giard 27).

The traditions of encyclopedism and Hermeticism come to the 16th century as tools for power in an economy of knowledge. Writers gained cultural capital through the circulation of their knowledge; readers gained capital by learning from the compiled knowledge. Books of secrets took their places with handbooks of practical knowledge as popular means for students to ensure career success in centralizing clerical and secular bureaucracies. But the manipulation of nature—what later will be called science and technology—was still seen as primarily a pursuit reserved for an initiated few who kept its secrets from the vulgar masses.

In the 16th century we can find books whose intent was to teach readers about particular subjects and these are still called "books of secrets," not textbooks. Paolo Rossi described the role of these books of secrets in the 16th century:

> The writing of these books was a response to a demand for practical information in a cheap and easily understandable form. They were an immediate success, and from 1530, there issued a stream of booklets on a wide variety of topics from astrology and veterinary medicine to

cooking and book-keeping. The important point about these works was the absence of technical jargon, and obscure philosophical rhetoric. They contained straightforward recipes which could be easily followed towards a practical end (170).

Books of secrets influenced 16th-century ideas about the relationship of science and the masses. Science had been carried out in secret by privileged groups up until this time. In the 16th century, however, popular clamor for books of secrets—and their consequent mass printings and distribution—created a tension between the people who sought to carry out scientific activities in private and those who sought to carry out these activities in public. Easy-to-follow recipes for manipulating nature, written in concise and clear language, enabled an increasingly public practice of science.

The Renaissance tension between private and public science had been playing out since the Middle Ages. For example, Augustine's doctrines against curiosity served to maintain religious authority in the face of pressure to legitimate knowledge gained by experience, such as the recipes for manipulating nature contained in the books of secrets. William Eamon presented another example of how doctrines against curiosity served to maintain the dominance of religious knowledge:

> Bernard [of Clairvaux (1090–1153)] wrote his admonition against "stepping out of one's station" with reference to Lucifer, "who fell from truth to curiosity when he turned his attention to something he coveted unlawfully and had the presumption to believe he could gain." But his real aim was to warn novices against prying into secrets that did not concern them, lest they upset the peace and stability of the monastic community. . . . The polemic against curiosity had weighty political overtones. For Bernard, humility was next to godliness and was a necessary condition of social order (66).

The religious community that sought to maintain its dominance in medieval European society saw threats to that dominance coming from the scholastics who held that nature could be known through logical speculation rather than theology. Scholastic science and curiosity threatened a social order based on religious authority.

For the scholastics who followed Aristotle's teachings, science was concerned with explaining the reasons for ordinary, natural events. It was not concerned with the extraordinary, which was deemed to be the province of magic and the books of secrets. Science was contained in scholarly texts and scientists developed theories and reasoning by consulting these texts, not by conducting experiments. Experimental—or experiential—knowledge was deemed the province of magic and was fit for the illiterate. In practice, scholars regularly carried out both textual and experimental work. But their textual science circulated in formal academic settings as legitimate, genuine knowledge while their experimental magic was reserved for private uses as non-legitimate, spurious knowledge.

THE UTILITY OF EXPERIENTIAL KNOWLEDGE

In the mid-16th century, Georgius Agricola published *De Re Metallica*, a compilation of knowledge about mining and metallurgy. In his introduction, Agricola cited the many textual and extra-textual sources that he consulted in compiling his book. Agricola himself was skilled in the craft about which he wrote and he supplemented his own knowledge by reading other authors' works and talking with people who were knowledgeable in mining and metallurgy. In his approach to compiling the knowledge included in *De Re Metallica*, Agricola was typical of other encyclopedists in the Greek and Roman tradition. Yet in other ways, Agricola was like the popularizers of secret lore (or magical knowledge) in 16th-century Europe. He presented recipes for manipulating nature, just as previous authors of books of secrets had done. He included information about alchemy and elves in the mines, which was the province of magic. But Agricola extended the encyclopedic and secrets traditions, synthesizing them with experimental knowledge in his text on mining that was to have far-reaching influences.

In *De Re Metallica*, Agricola sought to explain the reasoning behind some of the recipes for manipulating nature that he had taken from textual sources. Unlike traditional scholastic texts, though, Agricola included Hermetic texts as sources for his speculative reasoning and he treated what scholastics would consider occult topics, such as magnetism, whose workings could not be seen. For example, in Book II of his work, Agricola presented lengthy information about using a divining twig to locate veins of ore. But in addition to simply telling how to use the divining rod, he explained the reasoning behind the twig's movement by discussing magnetism and idiosyncratic human properties:

> when one of the miners or some other person holds the twig in his
> hands, and it is not turned by the force of a vein, this is due to some
> peculiarity of the individual, which hinders and impedes the power

of the vein, for since the power of the vein in turning and twisting the twig may be not unlike that of a magnet attracting and drawing iron toward itself, this hidden quality of a man weakens and breaks the force, just the same as garlic weakens and overcomes the strength of a magnet (39).

Here Agricola presented a recipe for using a divining rod based on occult magnetic principles, e.g., a "hidden quality of a man" can weaken or break the magnetic force; garlic weakens or breaks the magnetic force; a magnet draws iron toward itself; a person can use the magnetic force to find a vein of metal through the use of a divining rod. Agricola's work resembled the books of secrets in that his advice was in the form of a decontextualized recipe. Unlike books of secrets, however, Agricola related his recipe to Hermetic knowledge by discussing the place of occult knowledge in his 16th-century culture:

> Since this matter remains in dispute and causes much dissention amongst miners, I consider it ought to be examined on its own merits. The wizards . . . seek for veins with a diving rod shaped like a fork . . . it is not the form of the twig that matters, but the wizard's incantations which it would not become me to repeat, neither do I wish to do so. The Ancients . . . were also able to alter the forms of things by [the divining rod]; as when the magicians changed the rods of the Egyptians into serpents, as the writings of the Hebrews relate; and as in Homer, Minerva with a divining rod turned the aged Ulysses suddenly into a youth, and then restored him back again to old age; Circe also changed Ulysses' companions into beasts, but afterward gave them back again their human form; moreover by his rod, which was called 'Caduceus,' Mercury gave sleep to watchmen and awoke slumberers. Therefore it seems that the divining rod passed to the mines from its impure origin with the magicians. Then when good men shrank with horror from the incantations and rejected them, the twig was retained by the unsophisticated common miners, and in searching for new veins some traces of these ancient usages remain (40–41).

In this passage Agricola examined both the magical and practical uses of the divining rod, noting that the use of this occult instrument was in dispute "amongst miners." In this examination, Agricola sought to separate the magical history of the rod from its practical uses in mining. This separation was crucial to retaining the rod as an acceptable, respectable instrument of mining, since its magical history was caught in theological and social contests. In the 16th century, magic included all practices based on experiential knowledge that sought to manipulate nature. According to this formulation, Hermetic knowledge contained in books of secrets certainly was magical and knowledge about the physical world gained from practical experience—like that of mining and metallurgy—could also be considered magical. Thus, Agricola's entire subject matter in *De Re Metallica* verged on the magical. He definitely crossed over the line into magic, however, when he discussed occult subjects, such as magnetism and divining rods.

The problem Agricola faced in dealing with magical knowledge was that this type of knowledge bore the stigma of religious heresy—not to be taken lightly in the time of the Inquisition. Since the early sixth century after Augustine wrote *The City of God* (A.D. 426), magic was denounced as a form of idolatry because people practicing magic were in concert with demons.[2] According to this line of thinking, God intended nature's secrets to remain hidden from man. The story of Adam and Eve was an example of God's intended order of knowledge and secrets in the physical world, the violation of this order, and the tragic consequences of the violation. People who sought to manipulate nature through secret knowledge of magical practices threatened God's order. William Eamon cited Augustine's stricture on intellectual curiosity: "Augustine condemned magic as vain *curiositas* because it was a purposeless 'trying out' or 'tempting' (*tentatio*) of nature" (62). According to Bizzell and Herzberg, *The City of God* was written as a "comprehensive application of Christian ideas to the governance of the secular state" (382). Because this tract was concerned with social order, Augustine's stance on magic had social implications. As William Eamon argued, "Augustine's polemic against intellectual curiosity gained new relevance during the scholastic period, when reason reared its prideful head to challenge faith" (62). This challenge of magic and intellectual curiosity to religious authority and social order continued throughout the middle ages and into the 16th century when Agricola wrote *De Re Metallica*.

In Agricola's time, engineers and practitioners of the mechanical arts had access to books of secrets containing magical recipes. Since the purpose of the magical knowledge—the manipulation of nature for practical ends—coincided with the purpose of engineering and mechanical arts, engineers and craftsmen found these books useful. According to William Eamon,

> Medieval engineers enthusiastically appropriated magic as a theoretical framework for technology. Indeed they regarded magic as technology's sister art. Not only did learned magic give technology a

theoretical matrix, it served an important ideological function by promoting the image of the professional engineer as a magus who, with his inventions, manipulates nature's occult forces and gains mastery over the physical world. . . . [For some engineers,] the usefulness of the occult sciences in this world overcame any consternation about the dangers it may have held for the soul in the next (69–71).

Some engineers may not have been concerned with their souls, but evidently Agricola was. By separating the divining rod's magical history from its practical utility, Agricola first conceded that the divining rod had an "impure origin with the magicians" who consorted with demons and threatened the religious and social order. But when he "examined [the use of the rod] on its own merits," Agricola argued that miners who were "unsophisticated" in the ways of magic could still make use of the rod for locating veins of ore. In other words, the rod could be used to locate ore even though miners did not rely on magic to make it work. Agricola clearly stated his stand against magic when he declined to repeat "the wizard's incantations" that people sophisticated in magic would use to make the divining rod work. In accomplishing this separation of an occult natural phenomenon, such as the use of the divining rod, from the realm of magic, Agricola prepared the way for considering magnetism and other natural phenomena as legitimate objects of utilitarian scientific study.

Agricola's introduction to *De Re Metallica* also worked to differentiate his text from books of secrets. Instead of recounting how the information contained in the book was revealed to him in a personal encounter with a god—a generic literary device for giving books of secrets their authority—Agricola built the authority for his text on his own experience and that of people to whom he had talked and whose texts he had read. The presumption here was that the information contained in other people's texts was based on their experience and Agricola included only that textual material that seemed plausible to him in light of his own experience and reasoning. In this respect, Agricola's authority was built in the encyclopedic tradition, and although he included alchemical information from Hermetic texts, he also discussed the questionable nature of this information:

> Whether they can do these things or not I cannot decide; but, seeing that so many writers assure us with all earnestness that they have reached that goal for which they aimed, it would seem that faith might be placed in them, yet also seeing that we do not read of any of them ever having become rich by this art, nor do we now see them

growing rich, although so many nations everywhere have produced, and are producing, alchemists, and all of them are straining every nerve night and day to the end that they may heap a great quantity of gold and silver, I should say the matter is dubious (xxviii).

Agricola thus destabilized the information from books of secrets by undermining the authority of the books he cited. De Re Metallica resembled a book of secrets by addressing occult subject matter, but not by endorsing alchemy and magic. Agricola's final criterion for including occult knowledge was that he found it useful. In applying this experimental criterion and privileging firsthand information, while destabilizing the traditional repository for experimental knowledge (books of secrets), Agricola began to reconstruct the place of experimental knowledge based on its utility. In De Re Metallica, Agricola argued for valuing previously occult knowledge as genuine currency in a knowledge economy.

Agricola was one of many writers who compiled useful information into handbooks after the development of the printing press made these compilations profitable for press owners/publishers and writers.[3] With the popularization of this new type of handbook, many people could benefit from knowledge based both on ancient lore and personal experience. The experimental/experiential knowledge contained in these new handbooks, because it was separated from occult teachings, became more palatable and less heretical in a European culture influenced by Reformationist ideas of God and nature. In this setting, Francis Bacon brought together trends from the printing industry, popular handbook literature, religion, and statecraft to devise a plan for elevating the place of the mechanical arts within his culture and creating a popular project for using the mechanical arts for the betterment of living conditions. This plan has come down to the 20th century as inductive science. Bacon's full plan, however, was more comprehensive than simply a scientific method. It included social institutions and religious foundations, making science the vehicle for carrying out his social and religious project.

THE DEVELOPMENT OF PUBLIC SCIENCE

The contest for legitimation between scientific and non-scientific knowledge—between experimental observation and "armchair" speculation—can be traced to the works of Francis Bacon, who argued for a natural philosophy based on observation of the things of the world and against a philosophy based on logical speculation in the scholastic tradition. A case can be made for tracing this line of thought farther back in time to the works of Roger Bacon in the 13th century and William of Ockham in the 14th century.[4] I will trace this contest between scholastic and experiential or experimental knowledge, however, from Agricola's influences on Francis Bacon, since the latter Bacon is generally credited with articulating a more complete project for experimental science than were his predecessors. Furthermore, Benjamin Farrington argued that Francis Bacon's goals for science were different from those of earlier philosophers (32 ff.). Although his methods may resemble those of Roger Bacon and William of Ockham, his goal of experimental science as a public (not an elite) pursuit differentiate Francis Bacon's work from that of his English predecessors.

Some historians find that Bacon's project was a failure as good science because his ideas about inductive reasoning proved unworkable. For example, Robert Alexander Patterson credited Galileo with establishing the scientific method (152). J. G. Crowther, too, stated, "The method of research in physical science which has proved so successful during the last three centuries first appears in its complete form in Galileo's *Dialogues Concerning Two New Sciences*" (227). However, Crowther also compared Galileo's naivete concerning the politics surrounding his dealings with Pope Urban VIII and the Roman Inquisition to Bacon's political awareness:

> Harvey said that Bacon wrote on science like a lord chancellor. . . . If scientists are to save mankind and themselves now, they must no longer be pure Galileans, but become also Baconians, and remember that "knowledge, that tendeth but to satisfaction, is but as a courtesan," and its proper use is for "the benefit and relief of the state and society of man" (261).

Taking Crowther's point even further into a social/political context, Julian Martin made a strong case for the view that civil servant Bacon saw his work as being concerned more with politics, security, and social organization than with scientific experimentation or the betterment of the human condition. According to this line of thought, Bacon himself clearly placed his project for experimental science in a social context.

Francis Bacon read and was influenced by the 16th-century works of Georgius Agricola (Farrington 33 ff.). In these works, Agricola argued for a natural philosophy based on experience with and observation of the things of the world. For example, he wrote in *De Ortu et*

✳ *Causis Subterraneorum* (1546), "Those things which we see with our eyes and understand by means of our senses are more clearly to be demonstrated than if learned by means of reasoning" (qtd. in *De Re Metallica* xii). In other words, seekers of truth should experience and observe phenomena firsthand. Agricola's philosophic position embraced the idea that scientific knowledge could be gained more readily through practical experience and observation than through logic and speculation in the Aristotlean (or scholastic) tradition. Agricola's compilations of practical information represented a departure from the types of knowledge that had been legitimated through the dominant scholastic tradition, i.e., knowledge that tended to be speculative and based upon rigid tenets of Aristotlean syllogistic logic.

Bacon extended Agricola's thinking to include the study of all physical phenomena through a rational method of observation. Bacon also included a biblical rationale for repudiating Aristotle's authority. As other critics of the scholastic tradition had done,[5] Bacon charged that knowledge generated through logical speculation was impotent, unable to generate useful outcomes. Making knowledge in this scholastic tradition merely generated arguments about knowledge, he charged, but did not generate outcomes that made life better for people in general.

Bacon is generally credited in histories of Western thought with articulating an argument for scientific knowledge based on methodical observation of natural phenomena, with the "systematic reform of natural philosophy" (Creighton viii), or with seeking "to reform the whole of human learning in his time by proposing new methods for the development and communication of knowledge" (Zappen, "Bacon" 61). Bacon can also be seen as caught in a contest for knowledge legitimation waged between the dominant scholastics and the upstart Reformationists, adding a religious component to his scientific philosophy. Followers of the scholastic tradition and Aristotlean authority sought to maintain the authority of Christian Church rule.[6] This seeming paradox of upholding Church dominance through the authority of Aristotle's work can be understood in the context of what Edwin Hatch called the "metaphysical theology" (138) that resulted from using static Greek texts as the basis for Christian Church authority in the context of social and political change taking place in Renaissance Europe. This Aristotlean inspired metaphysical theology based its authority on claims of truth embodied in unchanging texts. These classical texts gained their authoritative truth value from being earlier in the course of human endeavor—closer to God's creation of the world—and, therefore, closer to God's wisdom than 17th-century writings.[7] Renaissance audiences considered contemporary texts to be corrupt by virtue of their temporal placement farther from the Creation within a declining human condition. Because the dominant Christian Church based its claim to social and political authority—as well as theological truth—on unchanging texts, the advent of printing and widespread distribution of texts posed a serious threat to the Church's secular and sacred authority.

The Church's attempts to maintain the security of its authority were contested by adversaries proposing other possible ways to order and understand the world. Some of these other ways of knowing the world would prove too strong for the dominant scholasticism and Church authority to resist. Knowledge legitimated through logical speculation would lose value in this contest; knowledge legitimated by what would become known as the "scientific method" would gain value because this type of knowledge promised more benefits for people's lives than scholastic knowledge. Rationalism, empiricism, objectivity, and science would gain social dominance through this Renaissance struggle with the Christian Church and scholasticism.

Many changes within the cultures of Renaissance Europe fueled the contest between scholasticism and science. Among these changes were the rise of nation-states, the decline of feudal governance, the development of gunpowder and a professional military, urbanization, the development of a mercantile class and currency. In addition, the development of new metal alloys enabled another Renaissance innovation—the printing press—to produce thousands of imprints from each printing plate and reproduce graphic images—feats not possible with wooden or soft metal printing plates. These improved printing presses produced millions of texts that flooded Renaissance Europe with unorthodox knowledge. Concurrently, explorers sailed to America and other far-off lands with newly developed compasses that enabled captains to sail ships well beyond sight of the coastline. These ships returned to Europe with artifacts from "primitive" lands, which the Europeans sought to order and understand within their cultural framework. In this context of a Renaissance information explosion, an alternative line of thought would come to the fore to deal methodically with this increasing body of information and challenge the dominant scholasticism upholding Church authority. This new line of thought is exemplified by the work of Francis Bacon, who in one sense sought to free "truth" from the hold of the Christian Church and a scholastic tradition based on Aristotlean logic. Of course, when Bacon contested the legitimated knowledge base of the Christian church, he necessarily contested the Church's dominant position in social and political spheres, as well as this institution's theology.

While Bacon set out a project for a methodical practice of science based on biblical authority, he simultaneously set out a project for social authority located in a secular state, adding a political component to his scientific philosophy. In contesting the knowledge and authority of the dominant Christian Church, Bacon conflated science and religion, moving the locus of knowledge to science/religion and the locus of authority to the secular state. P. M. Rattansi argued that Bacon made this move to secure what he saw as an unstable society beset by rapid change: "Bacon reflected a widespread conviction that the restoration of stability and order was necessary no less in intellectual matters than in the social, political, and religious spheres where a century of swift change had brought about a series of crises" (15). In Bacon's project, science/religion and the secular state worked in tandem to further a project for economic wealth, independence and increased control of the English Crown over that nation's governance—a project he had inherited from his father and uncles who were loyal servants of the Tudor dynasty before him.[8]

According to Bacon, scholasticism based on Aristotle's teachings was unable to effect positive changes in the human condition. Instead of communing with the things of the world, scholasticism occupied itself with "admiring and applauding the false powers of the mind" (*Instauration* 2). Bacon described Aristotle's method in the fragmentary *Thoughts on Human Knowledge* (1604):

> Thus it came about that Aristotle with his impetuous and overbearing wit was a sufficient authority to himself, despised his predecessors, compelled experience to lend a servile support to his own views and led her about like his captive slave . . . he shielded himself behind his Logic, alleging with more licence than truth that he had invented it himself, forced facts to dance attendance on his words, and by the self-seeking and cunning use to which he turned it corrupted the great variety of his learning and knowledge (Farrington 42).

Bacon set his project as overthrowing this scholastic concern with "false powers of the mind" and supplanting a useful pursuit of science based on the rational and methodical observation of natural phenomena in its stead:

> There can be but one course left, therefore,—to try the whole thing anew upon a better plan, and to commence a total reconstruction of sciences, arts, and all human knowledge, raised upon the proper foundations. And this, though in the project and undertaking it may seem a

thing infinite and beyond the powers of man, yet when it comes to be dealt with it will be found sound and sober, more so than what has been done hitherto. For of this there is some issue; whereas in what is now done in the matter of science there is only a whirling round about, and perpetual agitation, ending where it began (*Instauration* 2).

Bacon further described his mission as

the collecting and perfecting of a Natural and Experimental History, true and severe (unimcumbered with literature and book-learning), such as philosophy may be built upon . . . so at length, after the lapse of so many ages, philosophy and the sciences may no longer float in air, but rest on the solid foundation of experience of every kind, and the same well examined and weighed. I have provided the machine, but the stuff must be gathered from the facts of nature (*Instauration* 6).

By setting out a "machine" for the study of nature in *The Great Instauration*, Bacon separated the realms of the celestial and the terrestrial—the realms of the religious and the mundane—bringing science and philosophy into the terrestrial realm. To be more specific, Bacon brought science and philosophy in contact with the subterranean realms where ore is mined, as facts of nature are gathered in the quote above. These facts, like their analogous ore, can then be examined, weighed, and valued according to the assayer's estimate. Unlike celestial facts that are valued by a supreme and unknowable authority, terrestrial facts can be valued by human authority. By bringing science and philosophy out of the ether and into the world, Bacon rationalized the study of the low arts, such as mechanics, chemistry, mining, and metallurgy, based on their benefit to humankind.

Bacon's machine produced a social organization for public science that required many technical workers to accomplish the goals of the new natural philosophy. This social organization put forward by Bacon in the 17th century marked a radical break with traditional views of scientific practices as being protected in the domain of elite and cloistered groups. One of the ramifications of this call for a new social organization of science was that many people would have to be persuaded to take up Bacon's project of the "collecting and perfecting of a Natural and Experimental History," which he conceived to be a finite (if large) activity. James Zappen found that these histories were "the only part of the [scientific] method designed to permit general participation in science and its applications with an attitude of respect for nature and hope in the ultimate utility of the method" ("Historiography" 80). In other words, many people would have to take up scientific pursuits in order to accomplish Bacon's social project of improving the human condition through the methodical study of nature (and securing state governance for the Crown). Scientific knowledge would no longer belong to a cloistered elite; it would be public.

REFORMING SCHOLASTICISM

A major obstacle facing Bacon's project was the dominant scholastic tradition of dialectical instruction based on Aristotlean logic—a tradition in Church studies and authority that was already under attack in Bacon's time from English Reformationists. Bacon took up the Reformationist idea of man's ability to know the "truth" directly, an idea applied to religious truth in John Wycliffe's case and widened to include truths of natural philosophy by Bacon. In asserting that people could know truths directly, Bacon also asserted that people in his time did not know truths directly, that they had fallen from some purer state of direct knowledge, and were now engaged in corrupt relations with God and nature that needed to be reformed. That Bacon's project sought to restore humans to a god-like state after a period moral decay—to effect an instauration—is clearly seen in his subtitle to *The Masculine Birth of Time: Instauratio Magna Imperii Humani in Universum* (*The Great Restoration of the Power of Man over the Universe*). Bacon's project sought to restore humans to their pre-Fall condition described in the Book of Genesis, in which God promised domination of the world to Adam, Eve, and their progeny.

In his writings, Bacon was especially vehement in his opposition to what Farrington called the "triumph of Aristotle over St. Paul" (24), a "triumph" that Reformationists saw as having corrupted the moral fiber of Christianity by substituting dogma for evangelism. Bacon proposed to place practice again in the forefront of scientific work and Christian worship through his project for making useful knowledge through methodical observation of the things of this world. Or as James Creighton described it, "[H]is true mission was to recall men from the study of words to that of things, to point out to them the power and advantage to be gained from a true knowledge of nature" (iv). Bacon based his scientific project squarely on biblical teachings, which Farrington described in this way:

> . . . since God had made nature, the study of it was a religious duty, an act of worship, to be approached with humility and awe. Further,

science, like religion, fell under the law of love. There were virtually only two commandments—to worship God and love your neighbour. To love your neighbour meant actively doing him good. Science which, in one aspect, was the worship of God, was also, by the application of knowledge to the relief of man's distress, the means of realising the love of one's neighbour (28).

Seen in this way, Bacon's project for the methodical study of nature is also the worship of God because Bacon's goals were for "the benefit of the human race" (*Instauration* 3). He described his project as being

> not an opinion to be held, but a work to be done . . . I am labouring to lay the foundation, not of any sect or doctrine, but of human utility and power. Next, I ask [men] to deal fairly by their own interests, and . . . join in consultation for the common good; and . . . to come forward themselves and take part in that which remains to be done. . . . seeing also that it is by no means forgetful of the conditions of mortality and humanity (for it does not suppose that the work can be altogether completed within one generation, but provides for its being taken up by another); and finally that it seeks for the sciences not arrogantly in their little cells of human wit, but with reverence in the greater world (*Instauration* 16–17).

By placing his project within a biblical framework and emphasizing practical works over logical argumentation, Bacon set the groundwork for persuading other intellectuals, who were already influenced by Reformationist ideas, that scholastic speculation should be replaced by scientific observation. Because Bacon's scientific method was grounded in biblical teachings, experimenters could explore natural phenomena without fearing that they blasphemed by seeking to upset God's plan on earth. They could mine subterranean ores without fearing that they consorted with underworld demons. Unlike alchemists and practitioners of Hermetic secret knowledge, experimenters in Bacon's social organization could know the world through their senses because they followed God's word, which promised to improve the human condition through the domination of nature—a promise Bacon clearly grounded in biblical teachings in the "Preface" to *The History of the Winds* (1623):

> Wherefore if there be any humility towards the Creator, if there be any praise and reverence towards his works: if there be any charity towards men, and zeal to lessen human wants and sufferings; if there

be any love of truth in natural things, any hatred of darkness, any desire to purify the understanding; men are to be entreated again and
again that they should dismiss for a while, or at least put aside, those
inconstant and preposterous philosophies, which prefer theses to
hypothesis, have led experience captive, and triumphed over the
works of God; that they should humbly and with a certain reverence
draw near to the book of Creation; that they should there make a stay,
that on it they should meditate, and that then washed and clean they
should in chastity and integrity turn them from opinion (Spedding
3:207; translated in Farrington 54–55).

Following Bacon's advice to "draw near the book of Creation," in the
Book of Genesis God charged Adam and Eve to "'increase and multi-
ply, and fill the earth, and subdue it, and rule over the fishes of the sea
and the fowls of the air, and all living creatures that move upon the
earth" (1.28), a promise echoed in God's words to Noah after the flood.
In these charges, Bacon found a biblical basis for human domination
over Nature—not merely human service to or imitation of Nature. And
if Bacon's project proved successful, humans could eat and live without the "labour and toil" (Genesis 3.17) that befell Adam and Eve after
the Fall. This project, as Evelyn Fox Keller described it, would "transform not so much the world as [man's] relation to the world (35).
Bacon's science promised to transform the human condition into an
Eden-like state described in *New Atlantis*, where scientists were men of
God.

Bacon's social and religious project for public science was set out
between the lines of his writings, for the most part, and was concerned
with reforming and securing the governance of the state as much as
with improving the condition of its inhabitants. In this regard, Robert
Faulkner argued that "Bacon links a progressive state of scientific advance with other modes of controlling peoples" (232). In promising to
improve the human condition, Faulkner found that Bacon sought to
control people through hope and incentives [quoting from *De Augmen-
tis*]: "Bacon specializes in 'the politic and artificial nourishment of
hopes, and carrying men from hopes to hopes' . . . Addressing his
plans to the insecurities of life, Bacon, like Machiavelli, promises
progress in the means of security, but he can promise much more"
(201). It is an art of managing people that Bacon set out as much as a
project for dominating Nature. And by managing people through "artificial nourishment of hopes" Bacon sought to join the activities of natural philosophers and politicians into a new category of civil servant—
the practitioner of natural science who has theoretical knowledge that

places him above the technicians who carry out the social project of science. This theoretical knowledge, as Keller argued, would enable the natural scientist to understand and control Nature and society: "Science controls by following the dictates of nature, but these dictates include the requirement, even demand, for domination" (37). Science for Bacon, however, was more than a method for achieving human domination over Nature as promised in the Book of Creation. It was a method for achieving social dominance for science through the promise of increased security and an improved human condition. Technical language would become the lingua franca of this scientific society and its institutions.

3

The Rise of Experiential Knowledge and Technical Education

> *The ability to have made progress in empirical precision can hardly be denied to science. But the whole process of empirical classification, however precise and however well instrumented, will appear arbitrary, conventionally grounded in our concern to manipulate and control our environment, if it is not linked to rationally articulated theoretical conceptions.*
> —MARY TILES, *Bachelard: Science and Objectivity* (1984)

According to Bacon's project, the purpose of public science was to benefit humankind by manipulating nature and improving general living conditions. At the same time, science would enrich the rulers of nation-states and thereby solidify a centralized form of government. Theoreticians would plan the improvements, but the project was so vast that it would need a large number of trained, non-theoretical technicians to implement the improvements. This need for masses of workers in a public science was addressed in a number of ways: the sharing of knowledge through the proceedings of the Royal Society for the elite theoretical scientists, the sharing of empirical (what used to be called experimental) knowledge through how-to manuals for the masses and the elite, and institutionalized schooling to train masses of theoretical and non-theoretical workers. I shall address the latter two of these developments here in this discussion of textbook development.

Handbooks, such as Agricola's 16th-century work on mining and metallurgy, proved to be useful models for 17th-century how-to manuals focusing on the dissemination of practical knowledge for both workers, middle-class amateurs, and nobility who might dabble in the mechanical arts. One example of these how-to manuals was *Mechanick Exercises, or the Doctrine of Handy-Works*, written in the 1670s-80s by Joseph Moxon to teach the craft of smithing. Moxon originally published the manual in serial form, then compiled it into one volume published in 1687. In the third edition of this book, published in 1703, Moxon included an introduction in which he spelled out his reason for writing the manual:

> The Lord Bacon, in his Natural History, reckons that Philosophy would be improv'd, by having the Secrets of all Trades lye open; not only because much Experiemental Philosophy, is couch't amongst them; but also that the Trades themselves might, by a Philosopher, be improv'd. Besides, I find, that one Trade may borrow many Eminent Helps in Work of another Trade.
>
> Hiterto I cannot learn that any hath undertaken this Task, though I could have wisht it had been performed by an abler hand then mine; yet, since it is not, I have vetured [sic] upon it (n. pag.).

Moxon, who "enjoyed the friendship of England's brilliant scientists Robert Boyle, Edmund Halley, and Robert Hooke" (1), was a mathematician, publisher, and printer. He chose smithing as the topic of his manual because, according to his logic, this craft was the one upon which all other mechanical arts were based: "Besides, it is a great Introduction to most other Handy-Works, as Joynery, Turning, &c. they (with the Smith) working upon the Straight, Square, or Circle, though with different Tools, upon different Matter; and they all having dependance upon the Smith's Trade, and not the Smith upon them" (n.pag.). In starting public autodidactic education with the craft of smithing, Moxon addressed what he saw as the fundamental craft for those who would engage in the mechanical arts.

Bacon's project for a public science called for making knowledge through study of the mechanical arts and controlled experimentation, as opposed to the scholastic tradition of making scientific knowledge through rational speculation. In elevating the role of practical crafts and experimental knowledge from magic to science, Bacon took knowledge that had been considered the lore of illiterates and made it the subject of academic study. Accomplishing this cultural move called not only for autodidactic how-to manuals of mechanical arts, such as Moxon's, but also for a new system of schools in which empirical science could be taught. This new public academic institution needed a new kind of book to systematically train students in public science. Johann Comenius took up this part of Bacon's project.

Johann Comenius was a Bohemian Reformationist clergyman who lived most of his life in exile in 17th-century Europe. Comenius read and wrote widely, pursuing a pansophic ideal in his own work and in his educational philosophy. His educational philosophy was greatly influenced by Bacon's writings, particularly the *Magna Instauratio Scientiarum* and the *Novum Organum* (Jakubec 39). Educational historian F.V.N. Painter described Bacon's influence on Comenius' educational program:

> The scheme, suggested to [Comenius] by Bacon, was the publication of a work that would embrace and fully exhibit the whole circle of knowledge. This vast undertaking, which Comenius believed would be very helpful to the advancement of science, was obviously beyond the powers of any one man. Hence his practical mind suggested the establishment of an institution, in which all departments of learning should be represented by the ablest scholars, and from which this encyclopædia of knowledge was to proceed (204).

Comenius' plans for a "universal college" to complete his encyclopedia of knowledge was interrupted by the war between Ireland and England, so he traveled to Prussia and published his *Pansophia Diatyposis* (*The Great Didactic*) in 1643. In this tract, Comenius set out a plan for five graded levels of pansophic instruction: "[T]he doctrine of common concepts or ideas (idealia); natural science (naturalia) together with an account of man, his achievements and productions (artificialia); the doctrine of spiritual things in the Christian sense (spiritualia); and finally the section on how to know God and draw near to Him (aeterna)" (Jakubec 42). This idea of an institutionalized grading of education radically reconstructed the societal function of schooling.

In the Greek tradition, the goal of education up to Comenius' time had been to holistically mold the student as an upper-class citizen and shape him to participate ethically in civic activities. In Comenius' plan, large numbers of students, from all classes of society, would be educated through the technology of an educational institution. The goal of this instruction was not to create holistically prepared citizens. Rather, the goal of this institutionalized and disciplined education was to create trained workers who could carry out public science. The least trained of these workers would have completed the lower grades of Comenius' scheme (idealia, naturalia, artificialia). These lower grades would prepare students to carry out non-theoretical technician duties as described in Bacon's *New Atlantis*. The higher two grades (spiritualia and aeterna) would prepare the priests of science who would understand the theories behind technical work and would direct the lower classes of workers. Religious understanding was given to these higher order scientists, presumably so that they could maintain God's order while following their intellectual curiosity into the secrets of manipulating nature. This type of disciplined education—for large numbers of students across classes—called for a standardized approach to teaching, which in turn called for standardized textbooks to ensure that students moved systematically through the proper graded sequence of instruction.

In 1658, Comenius' textbook *Orbis pictus* was published in Amsterdam. This volume, which conformed to Comenius' reformist pedagogy as earlier set out in *The Great Didactic*, incorporated pictures and text in both Latin and a vernacular language to help students learn in a standardized sequence. In reconstructing the idea of books and learning in this systematized way, Comenius transformed handbooks, encyclopedias, manuals, books of secrets—earlier compilations of knowledge and/or information—into what we now recognize as textbooks: "*Orbis pictus* thereby initiated a tradition of school textbooks designed to be put in the hands of the children themselves" (Bowen III:104). *Orbis pictus* used the innovation of graphic illustration made possible by the printing press and copper plates, as had the authors of mechanical handbooks like Agricola and Moxon. Comenius combined this printing technology with a technology of teaching that he devised to train large numbers of students, from all classes of society, through standardized instruction. The textbooks that Comenius developed were designed for children to use in a graded and normalized educational institution, which would accomplish Bacon's (and Comenius') project of bettering the human condition and bringing people closer to God.

Comenius' first textbook offers a window for us to see how modern textbooks work to standardize instruction within disciplined educational institutions and social systems. *Orbis pictus* enabled students to learn in a graded sequence while it also normalized teachers' practices in their classrooms through the use of a textbook designed for both teachers and students. Up until Comenius' plan for institutionalized and graded instruction, students did not have textbooks themselves. Instead of reading for themselves from their own books, students received instruction orally from teachers who had the only books in the classroom. Teachers would read passages from great authors to the students, who might either recite these passages back orally or copy them onto slates. After paper became plentiful because of its use in printing presses, students might copy the passages into their own copybooks. Teachers might then lecture on the passage and dictate notes for the students to copy alongside the passages. With this type of teaching practice, the teacher—not the text—was the authority for knowledge in the classroom and what knowledge was taught depended on the teacher. This non-institutionalized teaching practice did not ensure that a standard curriculum was taught and served to destabilize the texts that were taught, since those texts were interpreted differently by each individual teacher. To further destabilize texts, both teachers and students could make mistakes in reading and copying the orally transmitted passages. This type of non-standard teaching practice was not well suited to the discipline of institutionalized education.

Because *Orbis pictus* could be used by both teachers and students in many classrooms, it taught graded content to students in a standardized manner. Now students and teachers had the same text in their hands. Students no longer needed to depend on the teacher for knowledge. The textbook contained the knowledge that would be transmitted in the classroom and students could read it for themselves. The technology of the textbook rivalled the teacher's firsthand experience as an authority in the classroom. Furthermore, this standard authoritative text was stabilized because each person in the classroom had the same text. Teachers no longer needed to read aloud for students to copy and errors were prevented. Textbooks also contained standardized commentary on passages from great authors. This system ensured that students learned a standardized curriculum, which in turn ensured that students were trained in a specific way when they left school. They could more easily enter the social institutions Bacon and Comenius envisioned for the betterment of living conditions through the manipulation of nature.

In addition to teaching standardized content to students, the textbook now taught teachers to teach in a standardized manner. Because the textbook was an authority in the classroom, teachers could look to it to provide them with appropriately graded levels of instruction. The textbook would know what information was appropriate for teaching common ideas to the newest students and how to build the higher levels of information onto this foundation at each successive grade. The textbook provided examples and commentary that would show the teacher how to work with the subject content at each graded level. The teacher only needed to consult the textbook to learn how this was done. Textbooks taught teachers how to teach within a disciplined educational institution. Teaching practice became more standardized after the 18th century to realize the social institutions that would improve living conditions through science.

JOHN LOCKE, LANGUAGE, PROPERTY RIGHTS, AND COINAGE

Bacon's influence on the institutionalization of science in 17th-century England is illustrated in the founding of the Royal Society, although Bacon's social plan was not adopted wholeheartedly in the Society's operations. The Society's conservative founders sought to neutralize forces, such as Agricola's use of alchemy and Bacon's plan for social reform, that they considered to be revolutionary and potentially destabilizing (Keller 47). The Royal Society's founders may not have entirely adopted Bacon's potentially revolutionary social project for public science, but both Spratt's and Andrade's histories of the Society noted Bacon's influences on Robert Boyle and other Society founders. For example, the arms granted to the Society with their Second Charter in 1663 bear the motto *Nullius in Verba*. Andrade explained this motto as announcing the Society's split from the scholastic tradition:

> . . . taken from Horace's
>
> Ac ne forte roges, quo me duce, quo lare tuter,
> Nullius addictus iurare in verba magistri.
>
> (And do not ask, by chance, what leader I follow or what godhead
> guards me. I am not bound to revere the word of any particular master.)
>
> This constituted a very clear indication that the Society cut loose from
> the authority of Aristotle and the other masters of antiquity (4).

Since this split from ancient authority was one of Bacon's enduring themes, the Society's motto suggested its debt to Bacon in its members' focus on gaining scientific knowledge through experience with the things of the world instead of through ancient texts. James Paradis, too, found that the Royal Society was indebted to Bacon for their approach to experimental science: "The motto of the Society, *Nullius in Verba* ('On the Word of No Man'), demanded physical proof as the condition of assent" ("Bacon" 208). Paradis further argued that the members of the Royal Society "viewed Bacon's *Novum Organum* as their essential text" ("Bacon" 207). Bacon's influence on the members of the Royal Society extended to their use of plain technical language as well as their empirical experimental methods.[1]

Francis Bacon's influences worked to refocus scientific inquiry during the 17th century onto practical studies that would generate useful knowledge for improving living conditions throughout society. His works also reformulated scientific knowledge as contingent, rather than being absolute as it had been formulated in the scholastic tradition. By comparing scientific inquiry to a hunt for concealed knowledge, Bacon again sought to replace scholastic inquiry with what he saw as a more utilitarian way of making knowledge. In viewing science as a hunt,

Bacon participated in a philosophical conversation that was to distinguish the study of philosophy from the study of theology—and the study of science from both of these. James Seth argued for the necessity of this distinction at this historical moment: "The necessity of differentiating ethics, as well as science and metaphysics, from theology was forced upon the modern mind by the dissolution of the politico-ecclesiastical system of the Middle Ages. The assertion of the independent authority of the State raised the question of the basis of its authority and the grounds of political obedience" (18). Seth further characterized the 17th-century philosophical conversation in England this way: "The first task of philosophy in the seventeenth century was to differentiate itself from theology, to assert the freedom of the scientific intellect from the bondage of authority, and to determine the proper method of this independent investigation of the nature of reality" (17). This conversation was extended by John Locke, and David Hume.

Like Bacon, Locke was concerned with practical knowledge. His approach to this knowledge was deeply influenced by scientific philosophy coming from the Royal Society, especially Locke's work with simple and complex ideas: "It must be remembered that Locke was explicitly imitating the scientists of the Royal Society in their methods. They were busy scrutinising natural phenomena and the behaviour of physical objects. Locke saw himself as doing the same for the mind" (Jenkins 39–40). That Locke's work was influenced by the Royal Society's empirical emphasis was a natural consequence of Locke's acquaintance with Robert Boyle. Locke historian Vere Chappell noted that Locke met Boyle in 1660 when Locke was a student at Oxford (6). As Jenkins pointed out, "Boyle was at the centre of the famous scientific circle that in 1663 became the Royal Society. The predominant feature of the new science, or the new 'natural philosophy' as it was then known, was its stress on practical research and experiment" (x). G. A. J. Rogers clearly explained Boyle's influence on Locke: "In Oxford, when [Locke] began serious scientific research it was as the assistant of Robert Boyle" (7). Peter Alexander, too, found that Locke took up Boyle's rejection of the idea of forces governing physical phenomena, in favor of an atomistic theory of matter: "Locke followed Boyle in favouring a corpuscularianism that avoided the use of forces on the ground that they are occult and not useful for explanation" (155). Locke's work under Boyle while a medical student at Oxford gave Locke ample opportunity to learn the new natural philosophy that sought to explain experience without resorting to occult, secret knowledge. The public knowledge of the new natural philosophy formed the basis of the

Royal Society's work for bettering living conditions through practical experiments.

Locke's practical philosophy addressed many topics that have important implications for technical communication, but three will be discussed briefly here: language, property rights, and coinage. Of language, Locke argued that science—and living conditions in general—would be more advanced if scientific knowledge was clearly communicated:

> it is ambition enough to be employed as an under-labourer in clearing ground a little, and removing some of the rubbish that lies in the way of knowledge; which certainly has been very much more advanced in the world, if the endeavours of ingenious and industrious men had not been much cumbered with the learned but frivolous use of uncouth, affected, or unintelligible terms introduced into the sciences, and there made an art of. . . . Vague and insignificant forms of speech, and abuse of language, have so long passed for mysteries of science; and hard or misapplied words, with little or no meaning, have, by prescription, such a right to be mistaken for deep learning and height of speculation; that it will not be easy to persuade either those who speak or those who hear them that they are but the covers of ignorance, and hinderance of true knowledge (*Essay,* "Epistle" xvi–xvii).

For Locke, clear language would allow scientific knowledge to flourish. Ridding scientific language of "vague and insignificant forms of speech" would serve to force out practitioners of secrets and mysteries; ridding it of meaningless words would force out speculation in favor of experimentation. The language of the new natural philosophy, according to Locke, was to be purified of the base usages of scholasticism and Hermeticism. This purified scientific language would lead to communication that would aid the "comfort and advantage of society" (*Essay* III. ii. 1: 323) through practical knowledge.

In *Two Treatises of Government* Locke explored what type of society could arise from the new natural philosophy of the Royal Society. In these treatises, Locke not only addressed concerns of the British state, but also set out a relationship between the Old World and the New World that could help to stabilize the British way of life. Historian Herman Lebovics argued, "Locke employed the vast unexploited resources of the New World to supply the key premise which lay at the foundation of the argument of his political philosophy. In his *Second Treatise* he summoned up the New World to validate the society emerging in the old" (567–68). In these treatises, Locke set out criteria for determining property rights in places that he considered to be in a "State of Nature" (*Second Treatise* 122 ff.) This discussion had both personal and national interests for Locke, who was

> secretary of the Lords Proprietors of Carolina (1668–71), secretary to the Council of Trade and Plantations (1673–4), and member of the Board of Trade (1696–1700) . . . one of six or eight men who closely invigilated and helped to shape the old colonial system during the Restoration. He invested in the slave-trading Royal Africa Trading Company (1671) and the Company of Merchant Adventurers to trade with the Bahamas (1672), and he was a Landgrave of the proprietary government of Carolina (Tully 168).

In his later life, much of Locke's income derived from investments in and dealings with the New World.

Among the many arguments that Locke made in the *Two Treatises* is one that justifies appropriating lands from indigenous peoples where they are living in a state of nature. According to this argument, settlers who cultivate and improve the land—thereby rendering the "greatest conveniences" from it—will have rights to the property:

> God gave the world to men in common; but since he gave it them for their benefit and the greatest conveniences of life they were capable to draw from it, it cannot be supposed he meant it should always remain common and uncultivated. He gave it to the use of the industrious and rational—and labour was to be his title to it (*Second Treatise* 137).

Historian John Tully argued that under this rationale, "Amerindians are then said to draw less than one-hundredth of the number of conveniences from the land that the English are able to produce" (183).

Failing to recognize the Amerindians replacement economy, British settlers under Locke's rationale could claim property rights because they took resources from the land. These resources could be used to create a favorable balance of trade for England, where Board of Trade member Locke saw excessive imports as a source of unstable coinage practices. Resources coming to England from colonies could correct this instability because those resources were part of England's domestic wealth. But while the settlers were claiming their property from the non-industrious Amerindians, those indigenous people experienced a different side of the coin:

In 1642, less than a quarter of a century after the coming of the Pilgrims, Miantonomo, a Narragansett sachem, charged: "Our father had plenty of deer and skins, our plains were full of deer, as also our woods, and of turkies, and our coves full of fish and fowl. But these English have gotten our land, they with scythes cut down the grass, and with axes fell the trees; their cows and horses eat the grass, and their hogs spoil our clam banks, and we shall all be starved (Pursell 14).

The Narragansett used natural resources in conservative ways that allowed those resources to be replaced and ensured that they maintained adequate supplies for Amerindian needs. According to Locke, however, people living in a state of nature, like the Narragansett, were not reaping as many conveniences from the land as could the industrious British settlers. The settlers, therefore, were justified in claiming the land as their own (under British rule) and substituting the Old World economy of excess production and trade for the Amerindian replacement economy. In this way, resources from the New World could enable the English economy to regain the favorable trade position and strength of currency that it enjoyed under Queen Elizabeth.

Locke addressed this currency question directly in *Several Papers Relating to Money, Interest and Trade, &c.* (1696). In these papers, Locke discussed at length why the value of silver coinage should not be raised. His argument rested primarily on the idea that if the value of coinage was raised by 20 percent, existing contracts would be paid at this devalued rate and landlords, for instance, would receive 20 percent less rent. This argument was valid only because Locke insisted that the value of silver coinage was in the quantity of silver contained in the coin, not in the king's stamp:

> *Raising* of Coin is but a specious word to deceive the unwary. It only gives the usual denomination of a greater quantity of Silver to the less . . . but adds no worth or real value to the Silver Coin, to make amends for its want of Silver. This is impossible to be done. For it is only the *quantity* of the Silver in it that is, and eternally will be, the measure of its value (*italics* substituted for Gothic type in original; "Further Considerations" 11).

And if the value of the coinage was in the silver alone, raising the value of the coinage would result in invalid contracts: " . . . the publick Authority is Guarantee for the performance of all legal Contracts. But Men are absolved from the performance of their legal contracts if the

quantity of Silver, under setled [sic] and legal denominations be al-
tered" (*italics* substituted for Gothic type in original; "Further Considerations" [9]). Thus, raising the value of silver coinage would destabilize the British economy and the king's rule with it.

For Locke, the sign of the coin was transparent. Foucault described
this way of viewing coinage and its relationship to attempts to standardize currency in the 16th century: " [M]oney does not truly measure unless its unit is a reality that really exists, to which any commodity whatever may be referred . . . arbitrary signs were not accorded the
value of real marks; money was a fair measure because it signified
nothing more than its power to standardize wealth on the basis of its
own material reality as wealth" (*Order* 169). When the sign for coinage
is transparent, wealth can only be increased by acquiring more metal.
In countries where silver is scarce, like England, there were two methods of acquiring precious metals: bullion must be exchanged for trade
goods exported in excess of imports; mines must be opened in other
countries. About this second option Locke wrote: "In Countries where
Domestick Mines do not supply it, nothing can bring in *Silver* but
Tribute or Trade. Tribute is the effect of Conquest.: [sic] Trade, of Skill
and Industry" (*italics* substituted for Gothic type in original; "Further
Considerations" 16). Again, Board of Trade member Locke advocated
claiming property rights based on either conquest or industry. To the
indigenous peoples living in states of nature in the New World, these
two options might have looked identical.

In Locke's philosophy, knowledge is contingent, worldly authority can be questioned, and science is limited to the study of mundane
things for the improvement of society in general. Language, in the
Royal Society tradition, was an invisible conduit for circulating scientific knowledge in an economy of utility, metals, and currency. Technical writing was the instrument for minting this currency of scientific
knowledge. Although Locke's purpose was to solidify a place for religious thought in philosophy, by separating science from religion he
freed scientific investigation and language from both philosophy and
religion. Hume took up Locke's attempts to secure a place for religion
in philosophy and extended the argument to a conclusion in which
the only possible knowledge was empirical. In Hume's skeptical
framework, science, religion, and philosophy were separate realms of
study, connected only by their dependence on conjectural, empirical
knowledge. Hume's philosophy, along with its forebears, impacts
technical writing practices through the works of Thomas H. Huxley—
that prolific and persuasive advocate of scientific and technical education.

DEFENDING SCIENCE AND
TECHNICAL EDUCATION

Huxley was familiar with the writings of Hume, having written a monograph on both Hume and Berkeley in 1878.[2] In an 1894 preface to this monograph, Huxley also noted his philosophic debt to Descartes:

> he, if any one, has a claim to the title of the father of modern philosophy. By this I mean that his general scheme of things, his conceptions of scientific method and of the conditions and limits of certainty, are far more essentially and characteristically modern than those of any of his immediate predecessors and successors (v).

Expanding on Descarte's idea of (un)certainty, Huxley cited Hume's contribution to the "reform of philosophy," stating that he "extended the Cartesian criticism to the whole range of propositions commonly 'taken for truth'; proved that, in a multitude of important instances, so far from possessing 'clear knowledge' that they may be so taken, we have none at all; and that our duty therefore is to remain silent; or to express, at most, suspended judgment" (x). For Huxley, certain knowledge was a rare commodity, gained only through firsthand observation and scientific proof—not based on *a priori* authority.

Scientific workers, in what Huxley termed the "modern spirit," worked tirelessly "gathering harvest after harvest of truth into its barns and devouring error with unquenchable fire" (x). In advising young English readers "animated by the much rarer desire for real knowledge," Huxley closed his preface by recommending the works of Berkeley, Hume, and Hobbes as sufficient for getting "a clear conception of the deepest problems set before the intellect of man" (xii). While Huxley acknowledged his philosophic debt to Descartes, he also felt that English readers could learn all they needed to know about scientific thought from the English philosophers.

Huxley also showed at least a passing familiarity with Bacon through his habit of quoting a passage from Bacon regularly in his public lectures. In "On Science and Art in Relation to Education" (1882), Huxley put forward one rendition of this passage: "A great lawyer-statesman and philosopher of a former age—I mean Francis Bacon—said that truth came out of error much more rapidly than it came out of confusion. There is a wonderful truth in that saying. Next to begin right in the world, the best of all things is to be clearly and definitely wrong" (173–74).[3] Huxley's repeated evocation of Bacon's philosophy suggests that Huxley had internalized at least some of Bacon's teachings, although he considered Bacon's method of scientific induction to be without merit.[4] Huxley believed that Bacon's project was a "magnificent failure" ("Progress of Science" 46) because it did not produce practical scientific results. But in his presidency of the Royal Society (1883–85) and his advocacy of institutionalized science and technical education, Huxley extended Bacon's legacy of public science for the betterment of society.

In his essay "The Progress of Science," Huxley reviewed fifty years of scientific achievements as part of a "Mr. T. Humphry Ward's book on *The Reign of Queen Victoria*, which was to celebrate the Jubilee year 1887" (*Life and Letters* II:129). Huxley began this essay by asserting that Bacon's plan for a public science "has proved hopelessly impracticable" (47) and finding fault with Bacon for not "winning solid material advantages" (47) from his scientific method. He then praised the fruits of Victorian science, while continuing to deny any connection to Bacon's work. In a passage describing "troops" of scientific workers, Huxley clearly tied scientific achievements to capitalist industry:

> And as the captains of industry have, at last, begun to be aware that the condition of success in that warfare, under the forms of peace, which is known as industrial competition, lies in the discipline of the troops and the use of arms of precision, just as much as it does in the warfare which is called war, their demand for that discipline, which is technical education, is reacting upon science in a manner which will, assuredly, stimulate its future growth to an incalculable extent (55).

On this side of the coin, industry provided science with the precise tools for advanced investigation. On the other side of the coin, science created "not 'new natures' in Bacon's sense, but a new Nature" through industry: "Every mechanical artifice, every chemically pure substance employed in manufacture, every abnormally fertile race of plants, or rapidly growing and fattening breed of animals, is a part of the new Nature created by science" (51). Huxley faulted Bacon for not producing scientific "fruits" in the 17th century. According to Huxley, the scientific "fruits" of the 19th century were not connected to Bacon's plan for a large number of workers working in the cause of public science for the betterment of living conditions. Yet Huxley celebrated these scientific "fruits" gained through the work of 19th-century "troops" of workers trained through technical education. Huxley had somehow differentiated Bacon's scientific workers in public science from his own troops of publicly trained scientific workers, thereby denying the roots of 19th-century science in 17th-century philosophy.

The apparent illogic of Huxley's argument in "The Progress of Science" might be explained by looking at how Huxley argued for separating what we now call "pure" science from "applied" science or technology. In reciting a roll call of scientists who were "guided by no search after practical fruits" (52), Huxley stipulated that true scientists were not concerned with practical applications of their work:

> In fact, the history of physical science teaches (and we cannot too carefully take the lesson to heart) that the practical advantages, attainable through its agency, never have been, and never will be, sufficiently attractive to men inspired by the inborn genius of the interpreter of Nature, to give them courage to undergo the toils and make the sacrifices which that calling requires from its votaries. That which stirs their pulses is the love of knowledge and the joy of discovery of the causes of things sung by the old poet—the supreme delight of extending the realm of law and order ever farther towards the unattainable goals of the infinitely great and the infinitely small (53).

While asserting that science is not beholden to industry, Huxley tried simultaneously to tie science to industry. After describing scientists here as religious geniuses—heirs to poetic realms—Huxley then described scientific workers as troops in war, which is a decidedly industrial (as well as national) pursuit. In making this early (illogical) argument for "pure" science and creating a role for scientists apart from the application of scientific knowledge to practical matters, Huxley could separate "pure" science from "applied" science or technology. Through this separation, Huxley could argue that "pure" science was unconcerned with practical social outcomes and, by logical extension, imply that "pure" science was not responsible for social outcomes. By separating "pure" science from technology, Huxley could make technology responsible for social outcomes while carving out a safe area for scientists to work in seeming autonomy. Scientific education could also be separate from technical education in this project and only technical workers need sully their hands with practical matters.

Nearing the end of a century that saw rapid and widespread industrial development in England, many intellectuals were beginning to question the desirability of their new industrialized lifestyles. Huxley, who worked in the tradition of practical philosophers such as Hobbes, Locke, and Hume, may have sought to differentiate and protect "pure" scientific work from the technical outcomes that he had championed so vigorously—technology that to some people seemed as menacing as it was promising.

As an interpreter of English philosophy, advocate of technical education, and university professor, Thomas Huxley had a lasting influence on the beginnings of technical writing as a disciplined practice in the United States. Not only did Huxley publicly speak on behalf of scientific and technical education in the English school curriculum, he spoke out against what he saw as a bankrupt curriculum built on the study of the classics—a debate reminiscent of Bacon's overthrow of the scholastic tradition in favor of scientific knowledge. In "Science and Culture" (1880), Huxley quoted from Matthew Arnold's *Essays in Criticism* (1865) both to agree with Arnold that culture is criticism and to disagree with him about how to attain that culture:

> Mr. Arnold tells us that the meaning of culture is "to know the best that has been thought and said in the world." It is the criticism of life contained in literature. That criticism regards "Europe as being, for intellectual and spiritual purposes, one great confederation . . . whose members have, for their common outfit, a knowledge of Greek, Roman, and Eastern antiquity, and of one another. Special, local, and temporary advantages being put out of account, that modern nation will in the intellectual and spiritual sphere make most progress, which most thoroughly carries out this programme. . . . "
>
> . . . I think we must all assent to the first proposition. For culture certainly means something quite different from learning or technical skill. . . .
>
> But we may agree to all this, and yet strongly dissent from the assumption that literature alone is competent to supply this knowledge.
> . . .
>
> Considering progress only in the "intellectual and spiritual sphere," I find myself wholly unable to admit that either nations or individuals will really advance, if their common outfit draws nothing from the stores of physical science. I should say that an army, without weapons of precision and with no particular base of operations, might more hopefully enter upon a campaign on the Rhine, than a man, devoid of a knowledge of what physical science has done in the last century, upon a criticism of life (142–44).

In this analogy of "criticism of life" as a military campaign, Huxley saw scientific knowledge as a precise weapon and science in general as the base of operations for waging successful social criticism. Therefore, preparing students for criticism entailed not Arnold's study of the classics alone, but Huxley's study of the physical sciences (physics and chemistry in particular), combined with the study of English literature, geometric drawing, writing (which for him is a kind of drawing ["Technical Education" 410]), and Latin, French, and German "because an enormous amount of anatomical knowledge is locked up in those languages" ("Technical Education" 409). For Huxley, technical education would be the weapon to save England from social unrest and upheaval.

Under the unchanging authority of classical authors, Huxley (like Bacon before him) felt that students were deprived of the firsthand experiences that would bring them true knowledge of the world. The literary education, in Huxley's terms, produced students who were not fit to win in the social contests that would prove English society either fit to survive or doomed to adaptation and/or extinction:

> Not only is [the student] devoid of all apprehension of scientific conceptions, not only does he fail to attach any meaning to the words "matter," "force," or "law" in their scientific senses, but, worse still, he has no notion of what it is to come into contact with Nature, or to lay his mind along-side of a physical fact, and try to conquer it, in the way our great naval hero told his captains to master their enemies. His whole mind has been given to books, and I am hardly exaggerating if I say that they are more real to him than Nature ("A Liberal Education" 116–17).

For Huxley, this matter of education was of vital importance for not only individual students, but for English society as a whole. Or as James Paradis argued, "The key to Huxley's political philosophy, if we may call it that, is education" (7). In speaking of his plan for technical education in the "Address on Behalf of the National Association for the Promotion of Technical Education" (1887), Huxley concluded by explaining his rationale for this type of education: "It is the need, while doing all these things, of keeping an eye, and an anxious eye, upon those measures which are necessary for the preservation of that stable and sound condition of the whole social organism which is the essential condition of real progress, and a chief end of all education" (445). For Huxley, education was the most likely remedy for the social unrest that England experienced in the late 19th century. It was also his remedy for maintaining a strong British colonial rule throughout the world, thus enabling English progress to continue through social evolution. Within the framework of social Darwinism, England was able to colonize less civilized peoples because English society was more highly evolved— and therefore stronger and more fit to survive—than the societies that were "in a state of nature," using Locke's terminology. Continued social evolution through technical education was necessary for England's continued prosperity through colonization.

Huxley set out his political philosophy in terms of the study of sociology, and rationalized his plan for technical education as a way to stabilize English society so it could progress. He saw English society as one of the greatest of all time, due in large part to the role of science in dominating nature for the betterment of all:

> The most thoroughly commercial people, the greatest voluntary wanderers and colonists the world has ever seen, are precisely the middle classes of this country. If there be a people which has been busy making history on the great scale for the last three hundred years . . . history which, if it happened to be that of Greece or Rome, we should study with avidity—it is the English. . . . If there be a nation whose prosperity depends absolutely and wholly upon their mastery over the forces of Nature, upon their intelligent apprehension of, and obedience to the laws of the creation and distribution of wealth, and of the stable equilibrium of the forces of society, it is precisely this nation ("A Liberal Education" 94–95).

In Huxley's plan, domestic stability within England and continued colonial rule of peoples living in "states of nature" depended on a sound scientific education for the "bringing of the mind directly into contact with fact, and practising the intellect in the completest form of induction; that is to say, in drawing conclusions from particular facts made known by immediate observation of Nature" ("Scientific Education" p. 126). Through this practice of inductive reasoning, students could learn not only how to think about the scientific and technological aspects of the work they would do, but they could also understand social phenomena correctly. In Huxley's view, technical education led to sound reasoning, social stability, a sound English economy based on colonial resources, and progress through the manipulation of nature.

Huxley had begun his teaching career in 1854 when he took a lectureship at the Government School of Mines at Jermyn Street. Thirty-one years later, Huxley was in ill health and decided to go into retirement. He resigned his position at the School of Mines and stepped down as president of the Royal Society. But he did retain a professorship in biology at the University of London, Royal College of Science: "He was asked to continue, as an Honorary Dean, a general supervision of the work he had done so much to organise, and he kept the title of Professor of Biology" (*Life and Letters* II:116). Huxley kept this post until the autumn of 1890 when he "returned temporarily to London to pack up his library, a considerable part of which he gave to the Royal College of Science" (Bibby 135). Sometime between 1885 and 1890 T. A. Rickard became one of Huxley's students. In a preface to his work *Man and Metals*, Rickard related how he and H. G. Wells were "fellow-student[s]" of Huxley: "It is a pleasing thought that the great teacher and expositor, Thomas Henry Huxley, under whose influence both of us were placed some forty-seven years ago, and ever since, would have commended highly the educative value of the work done by Mr. Wells in 'The Science of Life'" (vii). Since Rickard wrote a technical writing textbook in 1908 that, according to technical writing historian Robert J. Connors, "sold well and was adopted by a number of schools" ("The Rise" 332) in the United States, Rickard's debt to Huxley's philosophy has direct implications for the development of technical writing.

4

Contributing to a General Fund of Scientific Knowledge

The pioneer spirit is still vigorous within this Nation. Science offers a largely unexplored hinterland for the pioneer who has the tools for his task. The rewards of such exploration both for the Nation and the individual are great. Scientific progress is one essential key to our security as a nation, to our better health, to more jobs, to a higher standard of living, and to our cultural progress.

—VANNEVAR BUSH, *Science, the Endless Frontier*, 1945

According to the title pages of *A Guide to Technical Writing* and *Man and Metals*, Rickard was an editor of two engineering journals (*The Mining Magazine* published in London and *Engineering and Mining Journal* published in New York), as well as serving as an editor for the Mining and Scientific Press in San Francisco. Because Rickard had experience both as a practicing engineer and as an editor, he could speak about technical writing with expertise that was credible to other engineers.

In 1910, Rickard read a paper entitled "Standardization of English in Technical Literature" before the Institution of Mining and Metallurgy in London. In this paper, he set out the purpose of language and of technical writing:

> The purpose of language is to convey ideas; the intent of technical writing is to transmit accurate information, whether as fact or theory, from one man to another, to the gain of all. Indeed, the benefit is usually more to the giver than to the receiver. In the exchange of ideas, it is particularly true that it is more blessed to give than to receive. . . . At the start the writer finds his knowledge as full of holes as a sieve, and his thought as turbid as the pulp from a stamp-mill. In the effort to convey information by writing he crystallizes the amorphous ideas collected during years of study and observation, he submits

the confused notions in his brain to the settling process of logical thinking, whereby the true is precipitated from the false, the accurate decanted from the inaccurate, fact is filtered from supposition, and finally the solution of speech, pellucid but enriched, is outpoured generously.

The value of such a performance, either to the author or to his readers, depends upon the manner of it (*Guide* 128).

In this description of technical writing—and language in general— Rickard echoed some of Huxley's ideas about clear language as a conduit of objective meaning. In an unpublished paper, Huxley set out three conditions of good writing in praising the works of Defoe, Hobbes, and Gibbon:

> [B]y the dint of learning and thinking, they had acquired clear and vivid conceptions about one or other of the many aspects of man or things. In the second place, because they took infinite pains to embody these conceptions in language exactly adapted to convey them to other minds. In the third place, because they possessed that purely artistic sense of rhythm and proportion which enabled them to add grace to force, and, while loyal to the truth, make exactness subservient to beauty. . . . I have learned to spare no labour upon the process of acquiring clear ideas . . . and to regard rhetorical verbosity as the deadliest and most degrading of literary sins. Any one who possesses a tolerably clear head and a decent conscience should be able . . . thus to fulfill the first two conditions of good style. The carrying out of the third depends, neither on labour nor on honesty, but on that sense which is inborn in the literary artist (qtd. in Blinderman 53–54).

In their conceptions of language and technical writing, both Huxley and Rickard described language as a conduit for conveying ideas from one mind to another. Good writing, then, was judged by the effectiveness of this conveyance and in order to be effective, ideas should be conveyed in clear, vivid language that meant exactly what the writer intended. Through this effective conveyance of clear ideas, technical writers/engineers could each contribute what knowledge they possessed to what Rickard called a "general fund" of knowledge for the betterment of all. This "general fund" was crucial to humankind, because, according to Darwin's theory of evolution (popularized by Huxley and interpreted as social Darwinism by Huxley's friend Herbert Spencer), it is through the sharing of knowledge across time that humans evolve.

Rickard expressed the evolutionary importance of clear technical writing in a 1901 address to the American Association for the Advancement of Science entitled "A Plea for Greater Simplicity in the Language of Science":

> If the experience thus recorded were properly utilized . . . then man's advancement in knowledge and conduct would enable him to emphasize much more than is permitted him at the present, his superiority over the dumb brutes. Considered from this standpoint, language is a factor in the evolution of the race and an instrument that works for ethical progress. It is a gift most truly divine, which should be cherished as the ladder that has permitted of an ascent from the most humble beginnings and leads to the heights of a loftier destiny (*Guide* 125).

From this late-19th-century point of view, language was instrumental in the evolutionary development of humankind. This evolutionary and material point of view, termed "agnosticism" by Huxley, was central to a cultural state of mind in which the authority of religion was being replaced by the authority of rationalism and empirical science. Thus, Rickard described language and technical writing with religious overtones, such as "a gift most divine" and as a "ladder" permitting "ascent . . . to the heights of a loftier destiny." Rickard ended his talk by describing this "loftier destiny" as a state in which the writer "shall be linked to his fellow by the completeness of a perfect communion of ideas" (*Guide* 125). Man has become godlike through science and science is conveyed through clear technical writing. No wonder Huxley claimed, "[T]here can be no question that the magnitude and importance of the scientific literature of the present day has no precedent in human history" (qtd. in Blinderman 53). The very survival and development of the human race depended on science and technical writing, since the authority of God and religious teachings was brought into doubt by evolutionary theories of creation.

Against this backdrop of technical writing's importance to evolutionary development, Rickard discussed technical language as the coinage of intellectual and scientific exchange. In the same 1901 paper in which he described technical writing in terms of its value as coinage, Rickard also described technical writing in religious terms:

> The growth of knowledge has required an increase in the medium of intellectual exchange. . . . Sir Courtenay Boyle has pointed out that the purity of a nation's coinage is properly safeguarded, while the verbal

coinage of its national language is subject to no control. . . . The nation debases its language with slang, with hybrid and foreign words, the impure alloys and cheap imports of its verbal coinage, mere tokens that should not be legal tender on the intellectual exchanges (111–12).

In Rickard's view, scientific and technical language should be purified. Unlike Huxley, who felt that "art and literature and science are one" (qtd. in Blinderman 54), Rickard felt that "we lose such precision by making technical words do the chores of literary work" (*Guide* 45). Rickard set out a mechanical aesthetic of efficiency and purity for technical language, equating this purified technical language to the refined ore from which genuine coinage could be minted.

In Rickard's technical writing textbook, the first chapter after the introduction is entitled "Spurious Coin." In this chapter, Rickard discussed the debasing of technical language in mining and metallurgy through the adoption of foreign and "vulgar" terms:

> linguistic evolution advances in part, at least, by the adoption of words of lowly birth or even of those of illegitimate origin, but, if the exception be granted, there remains scant excuse for the employment of terms that come from the uneducated, seeing that we have the choice of synonyms that are the gift of scholars. . . . To some among us the crudities of speech heard in the mine and mill savor of the practical, and the exactness of the lecture room is suggestive of the theorist who does not soil his hands with labor or his clothes with grease. This is a pathetic fallacy (16).

Here Rickard argued to an audience of educated men that they are responsible for maintaining the purity of technical language. Unlike the people of lowly birth, these educated men could use more refined language and only chose to emulate lowly practitioners in an attempt to spurn their scholastic heritage. Like Bacon in the 17th century, Rickard participated in a contest between experiential and textual learning. In this iteration of the contest, scientific knowledge based on experience was dominant and in spurning textual scholastic knowledge, engineers indicated that they wished to be identified with science, not scholarship. Rickard, though, argued that the engineers' stance ultimately prohibited them from contributing effectively to the general fund of knowledge that would permit humankind to continue up the evolutionary ladder to perfect communion. By debasing the coin of technical language, engineers threatened science and the survival of the species. Religion was economics carried out through the currency of science and technical writing. The stamp of science gave technical language its economic value and religious potential, just as the stamp of the King gave coinage its representational value as currency.

Rickard summed up the importance of the purity of the coin of science—technical language—at the end of his chapter "Spurious Coin":

> The English language is the common heritage of the people of not one mining district, nor one region, nor one country, nor one continent . . . it is the heritage of the race to which Britishers, Americans, Canadians, Australians, and Afrikanders all belong, and also of the various races that they have assimilated in the course of their effort to conquer nature the world over. The mere fact that a word is distinctively Western Australian or Californian, is peculiar to Michigan or New Zealand, is reason enough for rejecting it. Let us have a mintage that will pass current at full value throughout the English-speaking world; let it be the refined gold of human speech (19).

In this plea for pure technical language, Rickard introduced arguments that were to form the foundation for the discipline of technical communication: clear, concise exposition to convey accurate ideas; correct English as the lingua franca of science and technical communication; consideration of an audience and making information clear for that audience; meaning as residing outside language; technical communication as a conduit for scientific information. Along with these familiar arguments, Rickard also introduced ideas from 19th-century discussions about religion and evolution that continue to reside in the technical language we use at the end of the 20th century: science and technical writing as ladders to a higher human destiny; doubts concerning the authority of religious teachings; contentions that text-based scholastic and classical knowledge is not as legitimate as experiential, scientific knowledge. To these religious and scientific views of technical writing, Rickard added economic arguments from his own experience as a mining engineer: technical writing is the currency of scientific knowledge; the stamp of science renders writing valuable; technical language must remain pure to safeguard scientific knowledge as the universal standard of value; scientific and technical language can colonize other languages, claiming resources at will from those languages for the benefit and in the name of science.

In constructing technical language as the coin of science, Rickard had moved from Locke's transparent sign (the coin's value equals the value of its metal as a commodity) to the representational sign (the coin's value is determined by the King's stamp that marks its denomination). In *The Order of Things*, Foucault traced this transformation of the concept of wealth from a transparent to a representational sign, arguing that in the 16th century coinage was a transparent sign of wealth:

> The mark that distinguishes money, determines it, renders it certain and acceptable to all, is thus reversible and may be construed in either direction: it refers to a quantity of metal that is a constant measure . . . but it also refers to certain commodities, variable in quantity and price, called metals . . . Here, the monetary sign cannot define its exchange value, and can be established as a mark only on a metallic mass which in turn defines its value in the scale of other commodities (171–72).

In this view, articulated by Locke in 1695, the coin's value is determined by the amount of metal it contains. The King's stamp on the coin verifies the weight and purity of the metal according to public standards of exchange. The metal in the coin functions as simply another exchange commodity.

With the rise of mercantilism, coins became more than just precious metal, they became representations of their potential exchange value. According to Foucault,

> the circle of "preciousness" is broken with the coming of mercantilism, and wealth becomes whatever is the object of needs and desires; it is split into elements that can be substituted for one another by the interplay of the coinage that signifies them; and the reciprocal relations of money and wealth are established in the form of circulation and exchange (*Order* 175).

With mercantilism in the 17th century came more widespread use of credit and other forms of trade that relied on representations of exchange rather than on the exchange of goods alone. It was in this context that Locke defended the transparent sign against pressures to split the representation rendered by the King's stamp of denomination from its correlative quantity and purity of metal. But Locke's argument for the transparent sign was overcome by mercantile desires and the need for representational signs of wealth. Foucault described the transformation in this way:

> Whereas the Renaissance based the two *functions* of coinage (measurement and substitution) on the double nature of its intrinsic *character*

(the fact that it was precious), the seventeenth century turns the analysis upside down: it is the exchanging function that serves as a foundation of the other two characters (its ability to measure and its capacity to receive a price thus appearing as *qualities* deriving from that *function*) (Foucault's italics, 174).

Once coinage *represented* wealth instead of *being* wealth, other commodities could also represent wealth. In Rickard's "worldwide traffic in ideas" (*Guide* 130), technical language represented the wealth inherent in scientific knowledge.

In the worldwide traffic of ideas as a commodity, technical language is not valued because of its intrinsic beauty, as poetry might be valued for its aesthetics. Instead, technical language is valued because it is a conduit for scientific knowledge—it represents science. As long as technical language bears this representation (the stamp of science), and *only* this representation, it can have value. If technical language is debased by foreign terms or by its historical connections to the liberal arts and Hermetic secrets, however, it loses value and threatens the stability of science. In the same way as Francis Bacon's uncle Thomas Gresham saw bad money driving out good, Rickard saw bad technical language driving out good science. And just as Gresham persuaded Queen Elizabeth's government to re-standardize the silver content of its coins in the Great Recoinage of 1560, Rickard argued that scientists should standardize their technical language. Through this standardization, scientific ideas could be freely and fairly traded in worldwide exchange.

At the turn of the century when Rickard was writing, the United States was in the midst of a heated debate about using gold as the standard for our currency. Since 1793, the United States had been on a bi-metal standard, coining both silver and gold once a mint was set up after the Revolution. During the Civil War, however, banks and the United States Treasury suspended payments in metals and paper money was used exclusively. In 1873, Congress revised the coinage laws and Roy Jastram explained the results: "[T]he public was not familiar with the American silver dollar of which they had seen few. Thus when Congress in its codification of 1873 *omitted* the silver dollar in its listing of future coins, no public attention was aroused by the omission. The legal effect, however, was that the right of free coinage of silver at the Mint had been discontinued" (139). The gold standard was legislated by omission.

In Rickard's time, a large contingent of the Democratic Party had noticed the omission and were arguing strenuously for a return to the bimetallic standard. In the Democratic Convention of 1896, William

Jennings Bryan gave his "Cross of Gold" speech that drew on themes of coinage, nationalism, and religion to depict an American people still freeing themselves from British influence. Bryan said,

> It is the issue of 1776 over again. Our ancestors, when but three millions in number, had the courage to declare their political independence of every other nation; shall we, their descendants, when we have grown to seventy millions, declare that we are less independent than our forefathers? . . . instead of having a gold standard because England has, we will restore bimetallism and then let England have bimetalism because the United States has it. . . . Having behind us the producing masses of this nation and the world, supported by the commercial interests, the laboring interests, and the toilers everywhere, we will answer their demand for a gold standard by saying to them: You shall not press down upon the brow of labor this crown of thorns; you shall not crucify mankind upon a cross of gold (117–18).

Bryan and Rickard addressed concerns for mining, metals, currency, language, and nationalism in the context of turn-of-the-century debates about the relationship of the United States and England, about the basis of our currency, about the politics of capital and labor. Bryan argued for the populace when he advocated the free coinage of silver for expanded wealth and trade. Rickard argued for the populace, as well, when he advocated correct English as the lingua franca of free worldwide trade in scientific ideas. In this worldwide trade, practical U.S. engineers would best English scientists when it came to producing technology for export. The United States would reassert its freedom and strength.

Like Locke, Rickard saw a favorable balance of trade as the way to bring resources into the United States, thereby strengthening our economy and improving our living conditions through science and technology. To this end, Rickard described his ideal engineer in *A Guide to Technical Writing*. He was simple-spoken and unself-conscious. He was "acquainted with several languages and the master of at least one of them" (30). He had a "quiet command of English and a masterful use of it" (31). He was "unaffected," sincere, careful, and disdained ignorance (32). He used "clear-cut thinking" (51) and did not exaggerate childishly (53). This ideal technical writer was knowledgeable and sympathetic (131), had a keen wit and courageous manhood (115). He said he didn't know something if he didn't know it (56). He was accurate and economical with his language; he would not condone "a Mormonism of style" (56). He did not recognize class distinctions, but did recognize "differences in bank balances" (62). He was not a snob (62), but was a "true American" (64). This characterization of the technical writer that emerges from background description in Rickard's textbook depicts the ideal of a rational, sympathetic early-19th-century gentleman/writer/engineer.

Unlike Huxley's ideal scientist as described in "The Progress of Science," who was not concerned with "practical advantages" but would toil and sacrifice simply to interpret nature, Rickard's writer/engineer was a practical man of business. Unlike Huxley's scientist who worked to further England's colonial economy, Rickard's American engineer worked to strengthen the economy of a former English colony. Because Rickard painted this technical writer as a being concerned with practical affairs, he strengthened Huxley's distinction between practitioners of "pure" science and practitioners of technology. By distinguishing between science as a pursuit unconcerned with practical applications and engineering as a pursuit concerned with these practicalities, Huxley and Rickard gave us a technical writer who served science second-hand. And through this secondhand service, the technical writer/engineer protected science from close association with doubts about industrial living conditions that haunted Western culture from the mid-19th century and became stark realizations with the technological horror of World War I. Technicians and engineers were on the front lines of battles over the place of science and industry in the early 20th century. Scientists remained in the rear of the action, protecting their "pure" activities from practical consequences and debasing technical language practices.

KNOWLEDGE IN TEXTBOOKS

At the turn of the 20th century, Rickard had set in writing many of the tenets of technical writing that continue to appear in current technical writing textbooks. Beyond the literal meanings of the advice in these books, though, what *kind* of knowledge is included in technical writing textbooks? In general, textbooks contain compilations of information about a specific topic in the traditions of encyclopedism, Hermeticism, and public science. In other words, textbooks attempt to exhaustively compile important knowledge about a specific topic in the encyclopedic tradition. But they also attempt to include recipe-like accounts of how to make the textbook information useful in some practical way in the tradition of the Hermetic books of secrets. And in their structure, textbooks present standardized information about idealized, graded practices in the tradition of public science introduced by Bacon and applied to institutionalized educational by Comenius. To summarize, we could (over)simplify this characterization to say that textbooks contain knowledge that purports to be exhaustive, important, useful, standard-ized, idealized, for the public benefit, and encouraging of systematized social stability through science. Thus, textbooks instantiate stabilized knowledge in an economy of scientific knowledge.

Technical writing textbooks resemble other types of textbooks in that they contain compilations of important and useful information that depicts an idealized and standardized practice of technical writ-ing. These textbooks also contain implicit information about social as-sumptions surrounding technical writing practices and the academic discipline of technical writing. This implicit information points to the roles technical writing plays in contests for knowledge legitimation and in institutional relationships. A noteworthy attribute of technical writing textbooks published in the United States throughout the 20th century is the apologia with which virtually all begin. What follows is a sample of these defenses for the study of technical writing and some assumptions that they communicate about the role of technical writing in institutional relationships.

In the late 1960s, Kenneth Houp and Thomas Pearsall presented their defense of the study of technical writing as a fictional case study of a junior engineer at work:

> His first experience with on-the-job reporting taught Ted four unfor-gettable things: (1) even a junior engineer is not simply a fix-it man whose only product is a gadget that works; (2) things that go on in your head and hands are lost unless they are recorded; (3) repeating what you have thought and done is a recurring necessity; and (4) re-porting, strange and difficult as it may seem at first, is something that can be learned by anyone of reasonable intelligence and perseverance.
>
> Ted Freedman is now earning $4.25 per hour as a junior engineer, and he will be eligible for promotion the first of July or January after obtaining his B.S. in Electrical Engineering. He lives with his wife and three-month-old daughter in Woodside Trailer Park (7).

This case-study version of the apologia for technical writing relied on an implicit argument that communication skills—and the technical knowl-edge upon which these skills are based—can result in personal economic gain. Houp and Pearsall alluded to this argument by including informa-tion about the young engineer's current rate of pay ($4.25 per hour) and stating that he will be eligible for promotion after earning his engineer-ing degree. The authors further emphasized financial implications by

adding that Ted lives in a trailer with his wife and young daughter. Presumably, students and teachers reading this textbook would assume
that Ted needed to support his family financially because his wife did
not earn a wage and that they aspired to own a home.

Houp and Pearsall's apologia explicitly stated four reasons for
studying technical writing in their four lessons that the young engineer
had learned: an engineer is more than "a fix-it man," engineers need to
record what they do, engineers need to repeat what they do, and "anyone of reasonable intelligence and perseverance" can learn to write
technically. This line of reasoning assumed that technical writing students were primarily male engineering students who needed to do
well on their jobs so they could earn a sufficient living to support their
families and buy homes. If the student succeeded in his profession, he
would be able to participate in an consumer economy and work to stabilize that economy through his engineering efforts at production and
his family's consumption of goods and services. His knowledge and
skills in technical writing would enable him to build up credits in an
economy based on scientific knowledge, thereby reaping professional
and personal rewards.

In addition, Houp and Pearsall's argument assumed that technical
writing is more of a craft than an art, therefore, nearly anyone could
become a technical writer with practice. Because the textbook would
set out easily followed recipes for communicating ideas, even a novice
engineer could gain the skills to succeed in his profession. In the handbook tradition, this textbook would include all important information
for being skilled in technical communication. In the Hermetic tradition, the authors made an argument for engaging this material based
on its utility and promise of financial reward.

In the early 1990s, Leslie Olsen and Thomas Huckin clearly defended the study of technical writing to students who would not find
this area of study inherently useful:

> Scientists and engineers may be technically brilliant and creative, but
> unless they can convince coworkers, clients, and supervisors of their
> worth, their technical skills will be unnoticed, unappreciated, and un-
> used. . . . From this perspective, communication skills are not just
> handy; they are critical tools for success, even survival, in "real
> world" environments (3).

These authors went on to provide pages of additional information jus-
tifying the study of technical writing, including five tables of survey
results to support their assertions about the importance of technical

communication in "real world" environments. Their extended defense of technical writing appealed to students' interest in doing well on their jobs, stating that communication skills are "critical tools for success." This personal financial benefit argument is probably the strongest appeal for students who see education as preparing them for practical success in the workplace.

The acute nature of this personal gain argument was further enhanced by the phrase "survival" in "'real world' environments," which echoed evolutionary arguments about the survival of the fittest—this time in the "real world" of the workplace. The phrase "real world" also signaled to the student that the classroom environment is admittedly not "real," or surely is not as significant as the world of work. So Olsen and Huckin's defense of technical writing assumed that the study of non-scientific knowledge in the academy is less legitimate than other types of practices, such as engineering work outside the classroom. Because this lesson was contained in a textbook, it served to train teachers of technical writing as well as to teach students to write technically. It also worked to maintain the separation between people with valuable technical knowledge (real world engineers) and those without it (academicians)—between people eligible for rewards in the economy of scientific knowledge and those who were ineligible.

In John Lannon's sixth edition of *Technical Writing* in 1994, he combined the personal financial benefit argument with an argument that bad communication skills were costly to "American business and industry":

> Your value to any organization will depend on how clearly and persuasively you communicate. Many working professionals spend at least 40 percent of their time writing or dealing with someone else's writing. . . .
>
> Good writing gives you and your ideas *visibility* and *authority* within your organization. Bad writing, on the other hand, is not only useless to readers and politically damaging to the writer, but also expensive: The estimated cost of communication in American business and industry is more than $75 billion yearly (Lannon's italics, 4–5).

What Lannon added to the apologia was a commodification of the student-as-engineer in his phrase "Your value to the organization." More specifically, Lannon talked about the value of the worker's communication practices, thus commodifying the worker's activities in the organization and clearly placing these communication activities in an economy of scientific knowledge. Lannon attempted to estimate the value of bad communication practices in "American business and industry"—$75 billion each year. Students reading this apologia presumably would not want to be costly to American industry, nor would they want to be inefficient. They know that in order to climb the organizational ladder to authority and personal financial benefit, they would need to be efficient workers. According to Lannon, 40 percent of their efficiency can be attributed to good communication skills. In other words, 40 percent of their value was dependent on their technical communication knowledge.

These defenses of technical writing were written for students who the textbook authors clearly assumed would think that writing and communication knowledge were less valuable to their career than technical and scientific knowledge. They were also written for teachers who need to rationalize the study of technical writing to these students, providing teachers with an argument in defense of their discipline. But how did this assumption about the need to defend the study of technical communication arise in the students', teachers', and textbook authors' minds? A overly brief explanation of this assumption is that it reflects the value of scientific knowledge in our culture. Because scientific and technological knowledge is the dominant way of knowing the world for us, the value of scientific study is not questioned. What *is* questioned is the value of studying some other way of knowing the world, such as through writing, which has a long history of affiliation with the liberal arts and only a recent affiliation with the sciences. In an attempt to raise the value of studying writing in students' eyes, these textbook authors have connected it to science through this argument: Your scientific ideas will not be valuable if you cannot communicate them. In other words, your ideas cannot circulate as currency in an economy of scientific knowledge unless you can communicate them well. This pervasive need to justify the study of technical writing to students who value scientific knowledge testifies to ongoing conflicts inherent in our culture between liberal arts-based and science-based knowledges.

A second aspect of our culture is apparent in these apologias for technical writing—the relationship of knowledge to management and money. This relationship is measured by career success in an organization, i.e., the salary an entry-level employee (recent college graduate) can earn by being promoted within an organization because of his or her technical communication skills. This linking of knowledge to money through a management technology works to ensure that technical writing students conform to behaviors and attitudes resulting in efficiency and productivity within organizations that have evolved from the application of time management and assembly-line models of production. Unlike early technical writing texts that viewed technical writers as workers in the public good contributing "something to the general fund [of knowledge]" (Rickard, *Guide* 130), textbooks in the post–World War II era construct technical writers primarily as individuals concerned with their own personal gain.

The linking of technical writing and money has a long and direct history, since technical writing has close ties to mining, precious metals, and coinage. For example, Agricola wrote the following description of his plain style in the preface to *De Re Metallica*: "Since the art of mining does not lend itself to elegant language, these books of mine are correspondingly lacking in refinement of style" (xxxi). He also described the importance of the many illustrations that accompanied his text:

> Although I have not fulfilled the task which I have undertaken [to write of the art of mining in its entirety], on account of the great magnitude of the subject, I have, at all events, endeavoured to fulfill it, for I have devoted much labour and care, and have even gone to some expense upon it; for with regard to the veins, tools, vessels, sluices, machines, and furnaces, I have not only described them, but have also hired illustrators to delineate their forms, lest descriptions which are conveyed by words should either not be understood by men of our own times, or should cause difficulty to posterity (xxx).

Agricola not only articulated a simple description of how a plain style of (technical) writing and visual illustrations enhanced a reader's comprehension of technical material, he also explicitly linked technical knowledge to economic gain and power:

> These books, most illustrious Princes, are dedicated to you for many reasons, and, above all others, because metals have proved of the greatest value to you; for though your ancestors drew rich profits from the revenues of their vast and wealthy territories, and likewise from the taxes which were paid by the foreigners by way of toll and by the natives by way of tithes, yet they drew far richer profits from the mines (xxxi).

In Agricola's time, feudalism was giving way to national governments, growing cities provided nourishing venues for an expanding mercantile trade, payments in coin rather than goods were becoming more common, and monarchs worked to consolidate their sovereignty over all sectors of society. In this changing social system, Agricola acknowledged the role his technical text played in supporting the monarch's power and wealth. Agricola clearly placed the knowledge of mining that his text conveyed within what we would now call power and knowledge systems. He forecast our current economy of scientific knowledge in which technical writing mints the coin of the realm.

TECHNICAL WRITING PRACTICES
IN POWER AND KNOWLEDGE SYSTEMS

Power and knowledge systems work to to bring order to knowledge. This ordering is accomplished through what Lyotard described in *The Differend* as the legitimation of some kinds of knowledge and the marginalization or silencing of other kinds of knowledge. This process of knowledge legitimation can be seen as a process of conquering (de Certeau, *Writing of History* 14), in which some possible statements are committed to discourse as knowledge while other possible statements are not committed to discourse and are, therefore, silenced. In Foucault's terms, silenced or subjugated knowledges can be of two types: "erudite" knowledges "buried and disguised in a functionalist coherence or formal systemisation" or "naive" knowledges "disqualified as inadequate to their task or insufficiently elaborated" (*Power/ Knowledge* 81–82).

In *Man and Metals*, Rickard described how the erudite knowledge of medieval miners was subsumed by knowledge of statecraft that changed the makeup of medieval society. In 13th-century feudal Europe, miners were among the first groups of artisans to form associations or guilds (549 ff.). The miners' associations were some of the strongest of the guilds for a couple of reasons: (1) they dealt in valuable commodities such as salt, precious metals, and gems (Renard 58); and (2) the products of their labors were highly sought by the royalty of newly forming nation-states (Rickard 601). These national rulers so desired the products of mining that they freed miners from their feudal responsibilities to a regional lord, such as taxation and military service, and declared that miners and the land they mined were in the King's demesne. This free status gave miners a privileged position in an otherwise feudal society, based on their knowledge of mining and their ability to fatten the monarch's coffers. But the miners' obligations to national monarchs also helped to strengthen emerging nation-states, thereby weakening the same feudal society that gave rise to their status.

The strengthening of national governments spelled the decline of trade guilds and mining associations. In 13th-century feudal Europe, miners had originally banded together because their work could not be accomplished by an individual. It required crews of miners to complete the mining operation; miners thought of themselves as one of a group. As time passed, mining associations included people who contributed money or mercantile know-how instead of direct labor. Thus, mining associations were transformed from labor cooperatives into capitalist organizations. As the makeup of mining organizations became more like businesses and national monarchs worked to consolidate sovereignty under their centralized rule, mining associations were put in the position of having to safeguard miners' working conditions against industrial capitalists whose interests were primarily in maximizing productivity and profits. In 1525, the Peasants' War dealt a strong blow to the influence of guilds and brought an end to an era of freedom and privileged status for miners.

The role of mining knowledge in the transformation of European society from feudalism to nationalism illustrates how the knowledge of mining that was privileged in the 13th century participated in a transformation of the society nurturing that privilege. In the transformation, mining knowledge was subjugated to knowledges of finance and state rule. As the transformation of feudal society progressed, people with mining knowledge were no longer the privileged freemen of earlier centuries; they became human machinery in mines they no longer owned. So the miners' knowledge participated in an economy of knowledge that both rewarded them and found them ineligible for rewards. The knowledge of mining that was dominant in one century was conquered by another century's knowledge of finance and/or national government. The miners' knowledge became erudite, in Foucault's terms, and moved from the power base to the margins of its power/knowledge system by the 18th century.

5

Engineering Specialized Social Organizations

As the first industry to make extensive use of the corporate form, railroads became the center of a contest over the form and place of the modern corporation in an industrial democracy.
—GERALD BERK, *Alternative Tracks*, 1994

By the turn of the 20th century, scientists had come a long way toward securing a professional role for themselves and legitimacy for their knowledge within the cultures of England and the United States. No longer did scientists suffer the fate Thomas Huxley wrote of in an 1851 letter: "To attempt to live by any scientific pursuit is a farce. Nothing but what is absolutely practical will go down in England. A man of science may earn great distinction, but not bread" (*Life and Letters* I: 72). Huxley's life itself epitomized the professionalization of scientists that took place within the last half of the 19th century. By the turn of the 20th century, scientists' knowledge was considered to be the basis for civilization itself and all the labor-saving mechanisms that improved the general quality of life in Western countries.

Because scientific knowledge was accorded this central role in Western cultures, other groups of practitioners who aspired to professional standing sought to align their practice with the sciences, showing how their knowledge was based on scientific principles and knowledge. Engineering was one of these budding professions that sought to demonstrate how its practice took "pure" scientific knowledge and used it to make mundane objects that would improve the general quality of life. Because engineering practice produced things that would be used by large numbers of people—translating scientific knowledge from scientists to non-scientists—engineers relied on technical communication to convey scientific knowledge to non-scientists, as well as to deposit applied scientific knowledge into Rickard's "general fund" of knowledge for the benefit of all engineers. Technical communication, then, became the lingua franca of science and engineering.

As engineering practice evolved within the large, complex organizations that were developed to accommodate modern business practices and products during the last half of the 19th century, engineers were called upon to design social as well as mechanical systems to control production and operation. These designs for social control were termed "management systems" and were an important focus of engineering practice in the United States from the 1880s to the start of World War I. As engineers designed management systems to make workers as efficient as the machines with which they worked, they also designed intricate technical communication systems as the mechanism for effecting operations control for maximum efficiency. Michel Foucault described such control systems based on observation and communication in *Discipline and Punish*:

> As the machinery of production became larger and more complex, as the number of workers and the division of labour increased, supervision became ever more necessary and more difficult. It became a special function, which had nevertheless to form an integral part of the production process . . . But, although the workers preferred a framework of a guild type to this new régime of surveillance, the employers saw that it was indissociable from the system of industrial production, private property and profit (174–75).

Management systems for control and discipline worked to make an organization's production more efficient by measuring each worker's performance and comparing it to pre-established performance and quality standards. This type of measuring and comparing viewed workers as individual units of production, not as members of collectives such as crews, gangs, or guilds as they were formerly conceptualized. This individuating process, made possible through technical communication, had an impact on workers that fundamentally changed the nature of organizational life: "These mechanisms [of surveillance] can only be seen as unimportant if one forgets the role of this instrumentation, minor but flawless, in the progressive objectification and the ever more subtle partitioning of individual behavior" (Foucault 173). In the management systems designed by engineers in the late-19th century, technical writing was the instrument for measuring and comparing the performance of individual workers and specialized machines.

This function of measuring and comparing individual performance against standards for production and quality allowed systematized management to operate through constant examination of machines and workers. Constant examination for deviance from standards constituted what Foucault called the "normalizing gaze": "The examination combines the techniques of an observing hierarchy and those of a normalizing judgment. It is a normalizing gaze, a surveillance that makes it possible to qualify, to classify and to punish. It establishes over individuals a visibility through which one differentiates them and judges them" (184). Without such constant examination, the management system could not judge whether individual machines and workers deviated from standards. Without such measurement and comparison, deviance could not be identified and corrected. Without such constant examination and correction, operations could not be systematically controlled. Technical writing—recording and reporting observed behaviors—was the mechanism for documenting this constant examination and correction and, thereby, controlling the management system through control of individual machines and workers.

The technical writing that conveyed information about individual workers and machines served as the mechanism for what Foucault called a "panoptic" system of surveillance for a "disciplined society" (198):

> This enclosed, segmented space, observed at every point, in which the individuals are inserted in a fixed place, in which the slightest movements are supervised, in which all events are recorded, in which an uninterrupted work of writing links the centre and the periphery, in which power is exercised without division, according to a continuous hierarchical figure, in which each individual is constantly located, examined and distributed among the other living beings . . . all this constitutes a compact model of the disciplinary mechanism (*Discipline* 197).

This panopticism—functioning through technical writing—enabled "the penetration of regulation into even the smallest details of everyday life" (198) within the systematically managed operation. No longer were shop workers able to decide how their work was done; written operation standards dictated in detail the most efficient ways to do their work. No longer could managers plan operations based on their idiosyncratic judgments; production and quality data were entered into standardized calculations to determine maximum future operations. Technical writing's panoptic episteme worked within a system of "hierarchy, surveillance, observation, writing" (Foucault, *Discipline* 198). This surveillance system ensured that the minute details of an individual's performance were available for correcting deviant behavior and planning future operations.

In the modern business organization by the turn of the 20th century, writing was the conduit for recording minute observations concerning the efficiency of workers and machines in order to discipline and control the organization's operation. Writing conveyed performance data between levels of hierarchical organizations, served as the mechanism for comparing this data to performance standards, and was the conduit for disciplinary actions to correct abnormal behavior or laudatory actions to reward correct behavior. Workers were punished or rewarded, in other words, based on observations of their behavior and data that were recorded from those observations. Most obviously, foremen and other supervisors observed and recorded worker behavior. But to make the panoptic management system function efficiently, observations must transcend the formal chain of command. Workers must be motivated to observe and report on other workers' behavior to fill any gaps in the formal hierarchical surveillance. For workers to report on other workers, though, they must think of themselves as individuals unassociated with a working group. Further, workers must have internalized the control system and its standards for behavior before they can be motivated to report abnormal behaviors on the part of other workers. As Foucault described this "major effect of the Panopticon . . . the inmates should be caught up in a power situation of which they are themselves the bearers" (*Discipline* 201). Workers take on the responsibility for maintaining the surveillance system through which their behaviors are observed, regulated, recorded, and rewarded or punished. Constant external surveillance then becomes unnecessary in the smoothly running panoptic system. Accordingly, the management system works most efficiently when workers are individuated and when they have internalized the system sufficiently not only to discipline and control their own behavior, but also to observe and report other workers' abnormal behavior. In this way, workers themselves complete the surveillance needed for maximum efficiency within the management system controlled through technical communication.

By the turn of the century, this panoptic discipline facilitated by technical writing was fundamental to the success of management systems based on private property and profits. But some workers objected to the dehumanizing effects of the system's discipline and control. In order to neutralize worker's objections, engineer/managers described the system and its goal of efficiency as "natural," implying that workers were objecting to an inevitable relationship between humans and nature. One influential engineer and writer on the subject of efficiency and management, Harrington Emerson, began his book *Effi-*

ciency as a Basis for Operation and Wages with this assertion: "Nature's operations are characterized by marvelous efficiency and by lavish prodigality. Man is a child of Nature as to prodigality, but not as to efficiency" (3). With this assertion, Emerson constructed management systems as engineer-designed replicas of nature's manufacturing operations. Here nature has become an abundant production plant, giving her children examples of both efficiency and waste. According to this argument, humans have learned to be wasteful from nature's example, but have not yet learned to be efficient. Emerson's argument echoed rationales for property rights and human welfare set down earlier by Bacon, Locke, Huxley, and Rickard. Resting on these established arguments, Emerson's assertion that management systems were necessary for teaching efficiency also argued for the right of organizations using these systems to claim property from those who did not use it "efficiently," as Locke had done with the Amerindians. In addition, this property could be claimed on the basis of efficient operation adding to one society's value and, therefore, its power relative to other less affluent societies. In this competition for resources, the management system that could ensure efficient use of raw materials (ore, human labor, water, land, etc.) was justified through its production value and social utility.

Management systems, therefore, were simply extensions of nature's production plant designed to correct humankind's behavior for continued prosperity. By couching his argument in these terms, Emerson's assertion that management systems are "natural" serves to remove the systems from their historical contexts, claiming instead that the systems are universal. As Roland Barthes argued, "This myth of the human 'condition' rests on a very old mystification, which always consists in placing Nature at the bottom of History. Any classic humanism postulates that in scratching the history of men a little, in the relativity of their institutions or the superficial diversity of their skins . . . one very quickly reaches the solid rock of a universal human nature" (101).

For the engineer/managers in the at the turn of the 20th century, human nature was exemplified by the large, complex organizations through which workers made products to improve general living conditions and owners made profits by virtue of their ownership of property, machines, and workers' labor. Workers could participate in the general improvement of living conditions by purchasing the products that they helped to make, thereby ensuring a market for these products and profits for factory owners. By calling this production system—and the management control on which it relied—"natural," engineers short-circuited questions about the inevitability of the control systems and the surveillance through technical writing that made the production system efficient and profitable. By the turn of the 20th century, management control systems and the technical writing that made them work seemed like part of the natural landscape of an industrialized U.S. economy.

ENGINEERING AS AN APPLICATION
OF PURE SCIENTIFIC KNOWLEDGE

Huxley's "pure" scientist may have been removed from the practical activities of engineers, but Rickard's engineer/technical writer worked in the thick of mundane matters. Huxley disliked this distinction between "pure" and "applied" sciences—a distinction paradoxically furthered by his public speeches on behalf of technical education. In "Science and Culture," an 1880 address on the opening of Sir Josiah Mason's Science College in Birmingham, Huxley objected to the idea of a body of technical knowledge that could be taught as "applied" sciences, separate from "pure" science:

> I often wish that this phrase, "applied science," had never been invented. For it suggests that there is a sort of scientific knowledge of direct practical use, which can be studied apart from another sort of scientific knowledge, which is of no practical utility, and which is termed "pure science." But there is no more complete fallacy than this. What people call applied science is nothing but the application of pure science to particular classes of problems (155).

In arguing for pure science as the source of all scientific and technical knowledge, Huxley insisted on the dominance of scientific knowledge and the subservience of technical or engineering knowledge. More than arguing for scientific education, Huxley argued for the dominance of a particular knowledge derived from empirical science. Science historian Ronald Kline traced this contest for scientific dominance over engineering or technological knowledge from the 1880s, finding that engineers subordinated themselves to "pure" science to enhance their status as professionals:

> In the late nineteenth century presidents of engineering societies . . . argue[d] that their field was not merely an "art" but a "science" as well, a move that raised their occupational status above that of

craftsmen to that of a "learned profession." Prominent engineers in the twentieth century carried on this tradition, climbing on the bandwagon of the growing prestige of science when threatened with a perceived loss of status (to technicians) as large numbers of engineers began to work for big corporations (221).

By the early 20th century, engineers working in industry asserted their professional status by aligning themselves with "pure" science, assuming the practical motives that differentiated their "applied" activities from those of the "pure" scientists who merely followed truth, no matter where it led them.

The relationship of "pure" and "applied" science continued to be debated throughout the 20th century. In 1944, for example, Franklin Roosevelt commissioned a study to recommend the role of science in the postwar United States. In this study, entitled *Science, the Endless Frontier,* Director of the U.S. Office of Scientific Research and Development Vannevar Bush described the relationship of "pure" and "applied" science (substituting the term "basic" for "pure" science):

> Basic research leads to new knowledge. It provides scientific capital. It creates the fund from which the practical applications of knowledge must be drawn. New products and new processes do not appear full-grown. They are founded on new principles and new conceptions, which in turn are painstakingly developed by research in the purest realms of science (19).

Here Bush described the same "fund" of scientific knowledge that Rickard had described at the turn of the century in his technical writing textbook. This fund of knowledge was the depository for scientific capital, "painstakingly developed" by scientists working in the "purest realms of science." These pure scientists were not motivated by practical goals; they simply followed truth, no matter how arduous the journey into uncharted frontiers of knowledge. Engineers could then draw from this fund of pure scientific knowledge for developing practical conveniences to improve living conditions and further social organization. Massachusetts Institute of Technology professor Dugald Jackson described the relationship of engineering to social organization in a way that is reminiscent of Bacon's scientific utopia in *New Atlantis*:

> The practice of invention . . . is subsequent in time to the development of reasoning powers in man and coordinate with the application of mind to social organization. Social organization proceeds hand in

hand with scientific discovery and invention, although the former may at times seem less lively in its forward steps than the latter . . . as anthropologists seem to have convincingly shown, there has been a long time for the more mentally active men who hated inconvenience to exert themselves in invention . . . the inventors of these improvements used intelligence to save exertion of muscles (594).

This connection between engineering based on pure scientific knowledge and social organization was more than simply speculation by Professor Jackson in the 1930s. In modern organizations of the early 20th century, engineers realized social designs as frequently as they realized blueprints for industrial designs.

DESIGNING SYSTEMATIC ADMINISTRATION FOR CANALS AND RAILROADS

Some of the first management designs contemplated by U.S. engineers were concerned with the technical administration of canals and railroads in the early 19th century. At the turn of the 19th century, most transportation of passengers and goods within the United States took place on waterways—oceans, bays, and gulfs; natural inland rivers and lakes; or constructed inland canals. At this same time in England, engineers were designing ways to attach steam engines to wheeled carriages to create locomotives for transporting coal from the mines at Newcastle. Cultural historian Wolfgang Schivelbusch described the evolution of mining car rails into locomotive rails in the Newcastle region at the turn of the 19th century: "The land between the mines and the river Tyne is covered by a dense network of railways up to ten miles long—descendants of the rails used in mountain mine shafts since the Middle Ages. These railways are appendages of the mines and are used exclusively to move coal" (6). From their beginnings in coal transport, railroads would expand their function to include passenger transport. This connection between coal and passenger transportation is symbolized in the early trains on England's pioneer Stockton-Darlington Line, which opened in 1825. These trains, a series of coal and passengers in coal cars, clearly illustrated the railroad's history in mining.

By the late 1820s, England's network of freight and passenger railroads attracted attention from people across the Atlantic at the fledgling Baltimore & Ohio Railroad, which opened in 1830. In an anonymous review of a biography of early civil engineer George Whistler, the reviewer recounted the situation of U.S. railroad engineering in the late 1820s:

> In the early days of the railroads, there were no engineers, even of the kind into which Major Whistler and his professional contemporaries developed. There were only military engineers, and consequently when some of these were detailed for work which was not military, the term "civil engineer" had to be invented to distinguish those who were no longer military. In 1828, the Baltimore & Ohio Railroad made a special request upon the Army Department for the services of Lieut. Whistler, and he with several others were sent to England to examine the railroads in that country. He was afterwards engaged on various railroads and canals (52).

If railroads had their English roots in mining technology, they had their U.S. roots in military engineering and canal building. These roots would later lead to the development of improved iron and steel for building coal-burning steam engines and durable rolled steel rails, as well as improved management systems for organizing railroad machines and workers.

Business historian Alfred Chandler argued that the construction of the Erie Canal and railroads during the 1820s and 1830s led to the development of organizational systems for controlling technical construction work. The technical administration systems that resulted from these large-scale projects needed to be supplemented, however, by a system for organizing the operation of the canal and railroad network:

> Often the chief engineer who planned the work became the general superintendent. . . . Even on a railroad, which before the 1850s was rarely more than fifty or a hundred miles in length, the general superintendent had little difficulty in keeping close contact with the division engineers and all the employees carrying out the road's business. Such enterprises has as yet little need for a systematic organizational structure (21).

An engineer/general superintendent could manage the operation of a canal because the canal company did not own the vessels that used the canal. The superintendent only oversaw the operation and maintenance of the locks and collected tolls.

Railroad companies required a more complex organization than canals, though, because railroad companies owned not only the roads, but the rights-of-way adjacent to the roads, the engines that ran on the roads, and many of the cars that transported freight and passengers: "the operation of trains required careful managerial structures of a kind not yet seen in the United States" (Salsbury 38). Railroad companies were the first to integrate many related functions into one organization, thereby creating an opportunity for far-flung enterprises and a need for more systematic control of operations.

Alfred Chandler placed the moment of this change in U.S. business organizations in the 1850s, when railroads lengthened their lines from the East Coast to the Great Lakes. No longer could one engineer/general superintendent "keep in close contact" with all employees: "But with the completion of the great east-west trunk lines early in the 1850s administration became a full-time task in American business" (21). By the mid-1850s, this concern with administration led a general superintendent of the New York & Erie Railroad to set out a management system in his report to the president of that railroad. Superintendent Daniel McCallum was one of the first to articulate a complete system for efficiently managing long railroad operations in a military-like system of rigid, panoptic hierarchical discipline: "The enforcement of a rigid system of discipline in the government of works of great magnitude is indispensable to success. All subordinates should be accountable to, and *be directed by their immediate superiors only*" (McCallum's italics, 104).

The system that McCallum instituted—which was emulated in other large railroad companies—relied on systematic reports from line workers to inform supervisors of operational details, thereby substituting reports for the personal contact that was impossible in large organizations. These reports would "give to the . . . responsible head of the running department a complete daily history of details in all their minutiae" (102), the better to realize efficient and profitable operations. McCallum's system posited an ideal operation and then rooted out deviation for discipline and correction. This system relied on technical writing inasmuch as reports played a pivotal role in identifying deviation. The fourth of six principles McCallum set out for the system addressed the connection between reporting and discipline: "Great promptness in the report of all derelictions of duty, that evils may be at once corrected" (102). The sixth principle addressed the connection between the system and discipline: "The adoption of a system as a whole, which will not only enable the General Superintendent to detect errors immediately, but will also point out the delinquent" (102). In concluding a section of his report entitled "System of Reports and Checks," McCallum summed up the importance of information contained in reports: "[I]t is only in [information's] practical application in pointing out the neglect and mismanagement which prevail, thus enabling us to remedy the defect, that its real value consists" (107). In McCallum's early and influential managerial system, reports were implemented as the mechanism for control and discipline, enabling the system to function at an ideally efficient and profitable standard. Thus, the technical writing embodied in these reports enabled panoptic surveillance, comparisons of operations to standards, corrections, rewards, efficient operations, the accumulation of capital, and the betterment of the human condition.

Railroads represented a new type of organization in the United States—one that was large in scale, geographically dispersed, internally complex and specialized, and vertically integrated. In response to the need for greater control than could be exercised personally by an owner, staff members in these large and complex organizations created panoptic control systems that relied on technical writing to prescribe standard operations and pinpoint deficiencies and deviations from these standards. Railroad timetables and rule books, for example, prescribed operating standards for transporting freight and passengers along the roads. Manufacturing specifications prescribed standards for goods purchased from vendors or built in company-owned shops. Line workers' reports pinpointed operations that deviated from standards. The goal of this systematizing was to create efficient and profitable operations; standards of efficiency and profitability were controlled through technical writing. As other industries realized opportunities for efficiency and profitability through large-scale organizations, these management systems, which originated in the railroads, were implemented more widely by the late-19th century.

ENGINEERS BECOME MANAGERS
IN COMPLEX SOCIAL ORGANIZATIONS

From the 1880s foreward, engineers working in industry increasingly found themselves supervising craftsmen and technicians as part of their organizational duties. In their new managerial roles, engineers redesigned forms of social organization to facilitate increasingly specialized and subdivided tasks for workers and machines. These new panoptic social systems for controlling both specialized tasks and the overall operation of complex organizations relied on the writing and recording of technical information for further coordination and control of operations. Business historian Mariann Jelinek identified two trends in the development of management systems in the late 19th century and early 20th century:

The first is a continuing attempt to transcend dependence upon the skills, memory, or capacity of any single individual. By recording the specifics of a task, a given outcome could be replicated: built into a formal system, institutionalized and independent of the individual. Instead of depending on the individual to discover anew the steps to be carried out, instructions could be specified. . . .

The second trend . . . is an attempt to rise above the concrete details of the task to think about what is being done, rather than merely to do it. Until this logical shift is made, coordination, forecasting, and real control are impossible. . . . Thus the two trends of thinking about and systematizing, or recording to insure replicability, are interrelated (64–65).

Engineers with managerial responsibilities at the turn of the century designed social organizations to systematize both production and managerial tasks so these tasks would not depend on individual skills or talents, but could be replicated by other workers in the future. At the same time, these organizations combined contradictory interests in observing and recording the minute details of individual performance with an "attempt to rise above the concrete details of the task." In combining these contradictory interests, engineers designed social organizations that were alloys of base and valuable components, e.g., manual and brain labor; individual and system performance; behavior and records of behavior. In an Age of Steel, management engineers sought to make stronger systems by combining social elements. These systems would work to stabilize relations between the two contradictory elements while combining them to form a stronger material. Management engineers assumed that an alloyed social system based on mutual benefit for labor and capital might induce these two groups with competing interests to work together through shared benefits.

As early as the 1880s, engineers recognized the need for improved social systems to meet the production goals of complex industrial organizations and the safety needs of industrialized consumer societies. In 1881, American Society of Mechanical Engineers (ASME) President Robert Thurston declared, "We are now entering fairly into the 'steel age'" (431), thereby signaling the U.S. culture's move from a relatively simple iron age to the next more organizationally complex age of metals. Early in his presidential address to the society, Thurston described the use of iron and steel in the late-19th century: "We are everywhere giving up the use of that expensive and perishable material, wood, and the weak and brittle minerals, and are substituting for them iron and steel. Iron is slowly but steadily and inevitably being displaced by steel" (425). The manufacture and use of steel enabled more complex organizational structures, such as the geographically dispersed railroads that relied on rolled steel rails. These complex "steel-age" organizations called for more systematic control of materials and workers. Thurston described an early concern with systematic materials inspection, which he had noted since the 1870s:

> Perhaps the most important advance made in the use of materials in engineering has been the general introduction of systematic inspection, and of careful test of all materials used. Such inspection and test are now demanded by every well-drawn specification . . .
>
> The Pennsylvania Railroad Company, the Bethlehem Iron Company, and other well-managed establishments have even organized complete departments devoted to the examination and test of all materials offered them . . . So essential is that system found to be that I am frequently called upon to advise in regard to new 'laboratories' . . . This is not a mere matter of dollars and cents, however. Every engineer who has experienced the anxiety which comes of uncertainty in regard to the character of the material of a structure, in which a single defective piece may cause the destruction of the whole with enormous loss of time, money, and probably of life, will understand what good comes of a system of inspection and test that relieve both conscience and pocket (428–29).

In the new industrial "laboratories" Thurston described, workers could examine and measure materials, comparing them to written quality standards and determining their deviance from these standards. This comprehensive testing system, like the larger management system, combined contradictory motivations of "conscience and pocket" into an alloy designed to be stronger than its individual elements. If engineers designed only from their consciences, they would build structures that minimized human cost but maximized dollar cost. If they designed only from their employers' pockets, on the other hand, they would build structures that maximized human cost but minimized dollar cost. The materials testing laboratory combined these two contradictory interests in an effort to build on the strengths in both economies.

Thurston described an early materials inspection system that relied on written specifications and reports, as well as inspectors who were trained in the use of these technical reports. Unlike previous inspection practices that relied primarily on the judgment of shop foremen, this modern inspection system relied on disciplined textual practices instead of practices based on an individual's unwritten knowledge gained from personal experience. By basing the modern inspection system on technical writing, "well-managed" organizations such as Pennsylvania Railroad and Bethlehem Iron could transcend the boundaries of any individual's knowledge and local geography. In the complex modern organization, interchangeable workers who were not physically proximate could be controlled through technical writings.

For engineers such as Thurston, their work was not only to build material goods for humankind. Engineers, like Bacon's workers in the cause of public science, could bring moral and intellectual good through their material works.[1] Thurston explained this connection in his 1881 address:

> The accumulation of wealth depends upon our material progress, and constitutes the only means of securing a steady progression in civilization, of conferring upon the world the blessings of intellectual and moral advancement, and of comfort and healthful luxury . . .
>
> Now, all that can be done in this direction must be done principally by the mechanic and the engineer, and our immediate duty is to see to it that our children and our children's children shall have every opportunity to acquire that knowledge and that intellectual power, and to gain those means and powers of directing the forces of nature (450–51).

By the 1880s, Thurston was able to extol the place of the mechanical artisan in society without apology. In more than two centuries since Bacon had argued for the value of mechanical arts and experimental knowledge over scholastic speculation, scientific knowledge had become dominant over speculation. By aligning itself with science, engineering need not defend its place in society. In fact, based on Bacon's religious justification for science, 19th-century engineers could assert their roles in "intellectual and moral advancement" of society, as well as the "comfort and healthful luxury" brought to humankind through their manipulation of nature.

Like Huxley, Thurston advocated technical education for all children "with the object of giving, in the least time and at the least cost, a preparation for all the duties coming to the learner" (451). In Thurston's plan, efficient technical education would be integrated into the social fabric of the country:

> With trade schools in every town, technical schools in every city, colleges of science and the arts in every State, and with a great technical university as the centre for the whole system, we shall yet see all com-

bined in a social organization that shall insure to every one absolute freedom to learn and to labor in any department of industry (452).

If engineers as social organizers created complex panoptic industrial systems based on technical communication and worker training, they also created the complementary educational systems for preparing workers and engineer/managers for these systems. As discussed in Chapter 3, standardized curricula for these educational systems—textbooks and normalized teacher instruction—ensured the systematic control of training for future workers and engineer/managers being prepared for life in systematized industrial organizations. As technical writing enabled a managerial system that could discipline and reward workers who internalized the system, textbooks enabled an educational system that could discipline and reward students and teachers who internalized the system.

In 1899, Slater Lewis wrote a series of articles for the *Engineering Magazine* in which he extended the idea of industrial system beyond Thurston's concerns with production and safety. In the first of these articles, Lewis advocated systematizing financial accounting to assign overhead costs to the items an organization produced. For Lewis, "the question is wholly one of increased efficiency of organization" (60). He saw a need for improved organization as the complement to improved production:

> Until quite recently the attention of engineers was almost entirely concentrated upon [design and efficiency in the articles sought to be produced]. The commercial arm of the business and the drawing office have been the happy hunting ground of the principal intent upon progress, whilst the internal routine and administration of shops, save perhaps in matters of purely engineering practice, have been left in the hands of "practical" men, with whom rule of thumb is supreme, and who commonly regard with surprise and contempt any suggestion that their arrangements need constant watching and periodical revision; still more that it is possible to mould them on a well-directed theoretical plan, coupled with a system of record that enables their weak points to be detected and the results of modification instantly to be seen (59).

In Lewis' plan for systematic management, "constant watching and periodical revision" of workers' practice was made possible by "a system of records"—by technical writing. In describing an inefficient organization, he noted that this inferior organization would show "reliance on verbal instructions, lax discipline, and ancient and ineffective methods of marshalling, dimensioning, and gauging work" (62). Lewis' management system watching workers constantly in a panoptic distribution was made possible by technical writing. In this iteration of the management system, technical writing disciplined more than production and safety; it controlled economics and efficiency in a system that relied on individual production, examination, and correction. In its reliance on constant examination and correction, this system enabled more efficient operations than were possible when operations were controlled by shop workers who did not record their practices and were outside examination and correction. Systematized management allowed higher levels of production, paid regular wages for labor, and produced consumers for mass-produced goods. Yet systematized management also weakened workers' sense of belonging to a group, isolated these workers for the purposes of examination, and indoctrinated individual workers into an ideology of efficient production and consumption that they were expected to internalize in order to succeed.

By 1899, the role of technical writing in systematic management had expanded to the heart of the system—its capital. Technical writing was no longer simply a representation of science alone; it also represented industry based on private ownership of engineering applications from scientific knowledge. Lewis' statement about "modern organization" applies to technical writing, since the systematized organization was based on written records: "The real object of modern organization is to strengthen the administrative arm in its control of routine, and to keep it closely informed as to fluctuations in sectional and departmental efficiencies" (63). The role of technical writing in systematic management was to strengthen managers' control of routine operations and to inform managers when operations deviated from the routine. In a 1900 *Engineering Magazine* article, A. Hamilton Church extended Lewis' systematizing of economic management, terming it "administrative organization":

> The object of commercial, or as it might also be termed, the administrative organization . . . should be to collect knowledge of what is going forward, not merely qualitatively, but quantitatively; it should also provide the means of regulating as well as the means of recording. It is no mere matter of accountancy, although the element of cost is an important factor; nor is the matter merely a technical one. It is essentially a matter of administrative control that is in question, stretching through each department and regulating the healthy life of the whole (393).

Here Church distinguished between technical and administrative knowledge, arguing that administrative knowledge controlled the entire organization by controlling each department in the subdivided and specialized organization. Church also saw administrative knowledge as encompassing accounting knowledge, but going beyond that specialization. For Church, administrative knowledge enabled the technology of a well-designed organizational system, much as mechanical knowledge enabled the technology of manufacturing machinery: "A modern system of organization is a high-class machine tool. It can be done without, but not economically. That is all there is to it" *(Proper Distribution* 15). Technical language was the mechanism through which administrative and mechanical knowledge was transformed into technologies of social systems and machines. Administrative knowledge was still in the domain of the engineer, whose applied knowledge was rooted in pure science. The stamp of science, therefore, continued to render value to technical language, even when it was used in a management context.

ENGINEERING MANAGEMENT SYSTEMS

When engineers set their attention on designing management systems to improve the efficiency of large and complex organizations, they began by designing pay systems for strengthening workers' motivation to increase productivity. This pay function combined many of the concerns that controlled the management system: maximum production for factory owners' benefit, standardized wages for worker's benefit, accurate record-keeping for controlling wages. Engineers sought to design pay systems that, like the management system itself, were alloys of elements stronger in their combination than they were individually. In addition to combining contradictory labor and capital interests, pay systems could also combine the conscience and the pocket, as Thurston had found in his material testing laboratories. In pay systems, however, engineers' conscience was expressed in religious terms echoing Bacon's call to return to the domination of nature as set out in the Book of Genesis. Engineers offered workers what engineers and factory owners considered to be a fair wage, thereby fulfilling capital's moral obligation to provide for labor's welfare. In return for this provision, capital could dominate labor as another natural resource in an industrialized economy, thereby taking care of factory owners' pockets.

The development of the most useful alloyed pay system took place over decades during the late-19th and early-20th centuries. In the mid-19th century, workers were most often paid according to day rates or piece rates. Day rate pay was thought to lead to worker ineffi- ciency because workers were paid a set rate for each day no matter how much they produced. Piece rate pay was thought to lead to greater worker efficiency because workers were were rewarded for in- creased production and, therefore, often produced more units under piece rate than under day rates. Typically, factory owners responded to this increased production by cutting piece rates. After workers were penalized a few times for increased production, they refused to in- crease their production in the first place, thereby keeping piece rates as high as possible and producing the minimum necessary to earn a competitive wage.

Both day rate and piece rate systems of pay were (un)satisfactory for either labor or capital. Day rates were satisfactory for labor because workers could control their rate of work and were ensured a constant wage. Day rates were unsatisfactory for capital, however, because fac- tory owners felt they could not control the work done on the shop floors. Piece rates, on the other hand, were satisfactory for capital be- cause factory owners could control more effectively the rate of work on the shop floors. Piece rates were not satisfactory for labor when they were coupled with cuts in rates to penalize increased production. To reconcile these positions of labor and capital, engineers designed profit-sharing systems through which owners agreed to distribute a percentage of the organization's profits to the workers. Engineer Henry Towne felt that profit sharing was unfair to labor, however, be- cause profits were determined by organizational activities over which workers had no control. Instead, Towne presented a "gain-sharing" system to the American Society of Mechanical Engineers in 1888.[2]

Towne's system removed an individual's economic interests from those of the organization as a collective: "The right solution of this problem will manifestly consist in alloting to each member of the or- ganization an interest in that portion of the profit fund which is or may be affected by his individual efforts or skill, and in protecting this interest against diminution resulting from the errors of others, or from extraneous causes not under his control" (601–2). Further, Towne envi- sioned that gain-sharing would unite factory owners and individual workers in partnerships where each party would enjoy the rewards of mutual efforts, while also retaining their "equitable rights":

> Certainly the problem we are considering will be best solved if it can be
> so formulated that the element of gratuity or charity . . . can be elimi-
> nated, and that, as presented to the employee, it becomes an invitation

from the principal that they should enter into an industrial partnership,
wherein each will retain, unimpaired, his existing equitable rights, but
will share with the other the benefits, if any are realized, of certain new
contributions made by each to the common interest (602).

From Towne's point of view, engineer/managers were on the side of
capital in this "problem" and the goal was to motivate labor to "new
contributions" that capital would then share with them in part. But in
this "partnership," labor would not gain any of the "equitable rights"
of capital, like deciding how work was done on the shop floors. In return for workers' cooperation, increased efficiency, and increased production, factory owners would agree to split with individual workers a
"portion of the profit fund which is or may be affected by his individual efforts or skill." This pay system did not destabilize the relationship between capital and labor. Factory owners retained the privilege
of ultimate control over their organizations. Workers retained the privilege of working for factory owners and receiving pay for their labor.

    The crux of this "labor problem" was voiced in the discussion fol-
lowing Towne's presentation, during which audience member Professor Denton noted the shortcomings of piece rate systems and his hope
for improving relations between workers and factory owners:

> I believe that constant reduction in piece prices, and all the time get-
> ting from the workman more than he originally did, has resulted in a
> sour state of mind on the part of the latter . . . the question must arise
> in an establishment that is carrying on piece work . . . How can we at
> once get the workman to squeeze a little more out of himself and at
> the same time be good-natured in doing it? A method of doing that is
> represented certainly in this idea of gain or profit sharing. I believe
> that the fact that the workman sees in it something to encourage him
> to go beyond piece work is likely to bring out a much better state of
> feeling between employer and employee than existed on the piece
> work system (616).

The labor problem, as viewed from the engineer/manager's stand-
point, was one of improving the "state of feeling between employer
and employee," while at the same time "squeez[ing] a little more" out
of the workers. The factory owners wanted to engineer more productivity from the workers, just as they had been able to engineer more
productivity out of the specialized machines with which those workers
labored. But workers, unlike machines, could react to re-engineering
with "a sour state of mind" or with "good-natured" partnership. The

problem with labor, from the factory owners' and engineers' point of view, was to get them to be "good-natured" about their roles as human machinery in the factories.

In further discussion of Towne's paper, audience member Charles Parker voiced his belief that if capital's motives were just, labor would recognize that justice and go along with capital's systematic management:

> I have found that nearly all the difficulties and troubles that arise regarding a fair and just arrangement, you may say, have a starting point in the selfishness of human nature. But I can say from experience that I never yet have seen any reason to lose faith; that where a spirit of justice and fairness is used on the part of the manufacturer, sooner or later . . . the same spirit has been shown by the workmen (619).

From capital's point of view, factory owners and engineers could overcome the inherent "selfishness of human nature" by living up to their responsibilities as the arbiters of justice and fairness. If owners were just, they could show workers that fighting against capital's plans for systematic management was unjust and unfair. Sooner or later, workers would agree with the owner's point of view and see the fairness of capital's control of labor.

Parker went on to predict this acquiescence of labor to capital based on a Christian grounding of management control: "[T]here is a way out of a great many troubles, and it is not only a business-like way, but it is a Christian way; it brings into it some moral principles" (621–22). Engineers could claim the moral and religious high ground for capital in its conflict with labor because engineering knowledge is based on pure scientific knowledge, which has been tied to religious teachings since Bacon justified experimental science using God's edict, contained in the Book of Genesis, that man dominate nature. In the capitalist culture of the late-19th century United States, labor had become a natural resource for man/factory owner to dominate, just as he dominated the metals that were made into specialized machinery and the coal and water that fed steam boilers to run those machines. The problem with labor, though, was that it did not see the inevitability of this relationship in the same way that capital did.

The task for engineer/managers in this conflict between labor and capital was to design ways to stabilize capital's domination over labor. In the best cases, engineers sought this stabilization through the persuasion inherent in the promise of increased wages. Engineers' focus on economic persuasion can be understood as an outgrowth of their role within the labor/capital industrial system. By the late-19th century, engineers clearly recognized their fundamental concern with finances, as well as with mechanics and physics. Towne addressed engineers' concern with finances in a paper he presented to the ASME in 1886, entitled "The Engineer as an Economist." Towne later quoted from this paper in his "Foreword" to Frederick Taylor's 1911 edition of "Shop Management:" "The monogram of our national initials, which is the symbol for our monetary unit, the dollar, is almost as frequently conjoined to the figures of an engineer's calculations as are the symbols indicating feet, minutes, pounds, or gallons. The final issue of his work, in probably a majority of cases, resolves itself into a question of dollars and cents" (5–6). Towne gave added weight to his concern with the economics of engineering by joining it with nationalism. In likening the dollar sign to the initials for the United States and placing this sign among the other signs in an engineer's lexicon, Towne made economics just another branch of scientific knowledge and made the engineer an economic agent within a national economy of science.

ASME President Coleman Sellers reiterated Towne's joining of engineering and economics in an address to the society in 1886. This address was later quoted by ASME President Alexander Humpheys in his 1912 address to the society, thus indicating the persistence of this idea of engineer as economist within the common sense of engineering: "The engineer who counts cost as nothing as compared to the result, who holds himself above the consideration of dollars and cents, has missed his vocation" (611). This idea of engineer as economist placed the engineer/manager between labor and capital when dealing with the "labor problem." In considering the labor problem from another (silenced) point of view, labor might just as well have seen the conflict as being a "capital problem." From this point of view, the surveillance and searching of workers' homes by Ford Motor Company's Sociology Department representatives and beatings by Ford's Security Department personnel could be called problems with overzealous representatives of capital's interest.[3] Because engineer/managers described the problem as being with labor, however, they indicated their position as being on the (dominant) capital side of the conflict. The president of the Stevens Institute Alumni Association[4] stated this dominant point of view in no uncertain terms when addressing students in 1896: "The financial side of engineering is always most important. . . . [The young engineer] must always be subservient to those who represent the money invested in the enterprise" (qtd. in Hughes 245–46).

If engineers represented the "money invested in the enterprise" for which they worked, and if engineers sought to remedy the "labor problem" to increase the enterprise's efficiency of production, engineers needed to design an organization in which capital controlled how work was done on shop floors. This meant that engineers had to design a new relationship between labor and capital, in which capital controlled the physical details of how workers labored. The design for this new relationship relied on self-interest, as articulated by Towne in discussing his paper "Gain-Sharing" with other ASME engineers:

> In my own opinion the time is coming very rapidly when some readjustment of the relations of labor and capital has got to be made, not necessarily by reason of the demands of labor organizations, but simply, if we disregard all questions of philanthropy or sympathy, from motives of self-interest on the part of the employer. Some better method of bringing out of men the best that that is in them in doing their work must be adopted (618).

The "best" that was in labor, from Towne's point of view, was not a moral or spiritual element in human beings. The "best" in this context was the most work a man could do. The engineer's job was to design a relationship between labor and capital in which workers would do the most work they possibly could. This new relationship was to be based on the self-interest of factory owners and workers. But because factory owners were dominant in this relationship, they would have the supreme authority over workers' self-interest. Capital would control labor through systematic management. Engineers would design management systems to accomplish this control through technical communication.

If factory owners wanted to control workers' productivity, accurate records of shop and organization activity were needed. McCallum had noted the importance of records for controlling individual's performance in his 1856 report for the New York and Erie Railroad discussed earlier in this chapter. Towne, too, argued for the primacy of records within his management system: The "starting-point of the [gain-sharing] system is an accurate knowledge of the present cost of a product . . . Where no such record exists, however, the only safe method consists in devising and putting into action a system of accounts which will furnish the desired *data*" (Towne's italics, 603–4). Systematic management depended on records—on technical writing—to control individuals' behavior and financial operations within the organization. This dependency was highlighted in Frederick Taylor's system of scientific management and was extended to the control of production and management itself.

6

Technical Writing as Management System Control

The growth of civilization goes hand in hand with adventure of the mind, just as advance of science is a child of the spirit of inquiry. Thus civilization and science have a bond of romance, and engineering is an agent by which civilization profits from science. Civilization connotes harmonious coopera- tion of many human beings, and also mutually sympathetic, helpful and ele- vated relationships. Civilization expands with the engineering arts because the latter enable groups of people to become closely associated without sacri- ficing either convenience or major comforts.
—DUGALD C. JACKSON, *"The Origins of Engineering,"* 1933

Taylor first introduced his system of scientific management in an 1895 paper presented to the American Society of Mechanical Engineers, en- titled "A Piece-Rate System, Being a Step Toward Partial Solution of the Labor Problem." In his scientific approach, Taylor modified the piece rate system to pay "two different rates for the same job; a high price per piece, in case the work is finished in the shortest possible time and in perfect condition, and a low price, if it takes a longer time to do the job, or if there are any imperfections" (857). In order to know "the shortest possible time," however, trained engineer/managers in Taylor's system first had to break a job into its simple component ac- tions and accurately observe and record how much time it took a pro- ficient worker to accomplish the actions. With this "accurate knowl- edge" (857), other managers in the rate-fixing department could determine a fair rate of pay based on "exact observation" (858) of the work to be done. The result of this fair rate of pay was that "the men are treated with greater uniformity and justice, and respond by doing more and better work" (858). Workers were paid according to their in- dividual performance, which must be carefully recorded in detail. Taylor cited the benefits of scientific management as "attaining the maximum productivity of each machine and man," "automatically se-lect[ing] and attract[ing] the best men for each class of work," and "promot[ing] a most friendly feeling between the men and their em-ployers, and so render[ing] labor unions and strikes unnecessary" (858). In its careful recording of time/motion studies, timetables for simple actions, instructions for different jobs, pay rates, employee production and performance data, machine production data, financial data, etc., Taylor's scientific management system openly relied on technical communication to control the organization, the workers, and the managers themselves. In scientific management, "the clerk be-comes one of the most valuable agents of the company" (862).

Taylor built on this idea that the clerk was one of the "most valu-able agents in the company" in a 1903 paper presented to the ASME in which he expanded on his scientific management system. This paper was further expanded and published in 1911 as his most influential monograph, "Shop Management." In this monograph, Taylor ex-plained how clerical and managerial labor should gain dominance by separating from manual labor, and how the efficient functioning of all three depended on technical writing. First, Taylor separated the differ-ent categories of labor: "[T]he cost of production is lowered by separat-ing the work of planning and the brain work as much as possible from the manual labor. When this is done, however, it is evident that the brain workers must be given sufficient work to keep them fully busy all the time" (121). Taylor had earlier described how clerks should record the many details required to control the organization. He then detailed the work of the managers:

> [T]he manager should receive only condensed, summarized, and *in-variably* comparative reports, covering, however, all of the elements entering into the management, and even these summaries should all be carefully gone over by an assistant before they reach the manager, and have all of the exceptions to the past averages or to the standards pointed out, both the especially good and the especially bad excep-tions, thus giving him in a few minutes a full view of progress which is being made, or the reverse (Taylor's italics, 126).

The function of technical writing within this complex organization started with the collection of data from clerks and proceeded upward into the management through reports and summaries generated by "brain" workers at various levels of the system. The data were increas-ingly refined as they moved upward to the managers, so that at the top levels of the organization, managers need only take a few minutes to compare the trends of the organization, labor, capital, and management performance to the standards that were set for their performance.

At first glance, labor seemed not to contribute to this fund of data and communication circulating within the scientifically managed organization. Instead, labor was the object of much of these communications—being reported on, rather than reporting. But Taylor noted an "important exception to this rule":

> "[T]he planning room gives its orders and instruction to the men mainly in writing and of necessity must also receive prompt and reliable written returns and reports which shall enable its members to issue orders for the next movement of each piece, lay out the work for each man for the following day, properly post the balance of work and materials accounts, enter the record on cost accounts and also enter the time and pay of each man on the pay sheet. There is no question that all of this information can be given both better and cheaper by the workman direct than through the intermediary of a walking time keeper . . . the only way in which workmen can be induced to write out all of this information accurately and promptly is by having each man write his own time while on day work and pay when on piece work on the same card on which he is to enter the other desired information, and then refusing to enter his pay on the pay sheet until after all of the required information has been correctly given by him (127–28).

Under Taylor's system of scientific management, individual workers reported on themselves, thus ensuring that they internalized the system. The workers who could not or would not internalize the system were forced out through penalties such as withholding their pay in the above case, or receiving substantially lower rates of pay under the differential rate system. The workers who internalized the system and reported on themselves were rewarded by higher rates of pay and promotion to jobs higher in the organization. Among these successful workers, according to Taylor, the "feeling that substantial justice is being done them renders them on the whole much more manly, straight-forward, and truthful. They work more cheerfully, and are more obliging to one another and their employers" ("Piece Rate" 881). These successful workers disciplined themselves within the scientific system of management. Technical writing and reporting provided the mechanism for ensuring that workers internalized the system, just as it was the mechanism for management control of the organization.

Although technical writing controlled management systems, along with the people and machines working in these systems, it also stabilized the ultimate control of capital over labor within the system. As in Towne's earlier gain-sharing system of pay, Taylor's scientific management system relied on a "partnership" between workers and factory owners in which factory owners and their managerial agents controlled how work was done on shop floors. In return for relinquishing control of how they did their work, workers who internalized scientific management were rewarded with higher-than-average wages— as long as they produced higher-than-average rates and quality of work. But owners and their managerial agents determined work rates and quality standards, thereby controlling workers' wages as well. Taylor explained this stabilized relation between capital and labor from the owner's, the engineer's, the manager's point of view:

> And this means *high wages* and a *low labor cost*. These conditions not only serve the best interests of the employer, but they tend to raise each workman to the highest level which he is fitted to attain by making him use his best faculties, forcing him to become and remain ambitious and energetic, and giving him sufficient pay to live better than in the past (Taylor's italics, "Shop" 29).

Shortly before this passage in the early pages of "Shop Management," Taylor had likened workers to horses, thus setting the analogy for the passage just cited above:

> He would not for an instant advocate the use of a high-priced tradesman to do the work which could be done by a trained laborer or a lower-priced man. No one would think of using a fine trotter to draw a grocery wagon nor a Percheron to do the work of a little mule. No more should a mechanic be allowed to do work for which a trained laborer can be used (27).

From the owner/engineer/manager's point of view, labor was a natural resource to be dominated as man dominated horses. Workers were only so much brute power to deploy alongside machines to generate greater power than was formerly attainable with men, horses, and crude implements. Instead of driving horses in a pre-industrial system, men had become the horses in this new industrial system. Under scientific management, the brain-workers' job was to drive manual-workers as men formerly drove horses: "making [them] use [their] best faculties, forcing [them] to become and remain . . . energetic, and giving

[them] sufficient" sustenance to be better kept than the other horses in the neighborhood. Workers who internalized scientific management may have been better paid than their neighbors who worked for average day or piece rates, but they and their neighbors shared the same relationship with dominant capital. And as the ranks of brain-workers grew, managerial agents of factory owners also shared labor's relationship with dominant capital.

NATURAL AND MILITARY EFFICIENCY

Managers in Taylor's scientific system practiced what Taylor called an "art of management," with efficiency as their goal: "The art of management has been defined, 'as knowing exactly what you want men to do, and then seeing that they do it in the best and cheapest way'" ("Shop" 21). Managers in other management systems also pursued efficiency as their organizational goal, but systems differed in how they were structured for efficiency. Most management systems followed a military structure, which Taylor described in "Shop Management" using the example of "a large engineering establishment building miscellaneous machinery":

> Practically all of the shops of this class are organized upon what may be called the military plan. The orders from the general are transmitted through the colonels, majors, captains, lieutenants and non-commissioned officers to the men. In the same way the orders in industrial establishments go from the manager through superintendents, foremen of shops, assistant foremen and gang bosses to the men (92–93).

Taylor argued that this military organization was not suited to scientific management, since it called upon managers at each level of the hierarchy to carry out too many diverse duties. Taylor called for *"abandoning the military type of organization"* (Taylor's italics, "Shop" 98) and instituting "functional management," which he described this way: "'Functional management' consists in so dividing the work of management that each man from the assistant superintendent down shall have as few functions as possible to perform. If practicable the work of each man in the management should be confined to the performance of a single leading function" (99). As a result of functional management, a worker might "receive his daily orders and help directly from eight different bosses, each of whom performs his own particular function" (99).

With his idea of functional management, Taylor extended the trend toward specialization in machinery and applied it to management workers. Just as his system broke down manual work into its component activities and machine work into specialized production functions for time analysis and control, functional management broke down brain work into its simplest activities for specialized production.

Taylor's scientific management system was not widely adopted in its entirety, although Taylor's concerns with efficiency and the development of management as a science were widely shared by other engineers, managers, and factory owners.[1] Functional management was among the components of scientific management that was not widely adopted as Taylor had set it out. Instead, organizations developed an organizational model that combined military and functional management. The strengths of this alloyed military model for efficiency were explored in depth by engineer Harrington Emerson, who originally published a series of articles on this subject in the *Engineering Magazine* in the early 1900s. These early articles were compiled into a book first published in 1908, entitled *Efficiency as a Basis for Operation and Wages*, in which Emerson detailed his "Gospel of Efficiency" (241). He based his argument for efficiency in production on nature's example in the first sentences of Chapter I: "Man is a child of Nature as to prodigality, but not as to efficiency. If it had happened the other way—if he had followed Nature's lead as to efficiency, but had taken up parsimony as a distinctly human virtue—the human race would have been wealthy beyond conception" (3). In speaking of parsimony versus efficiency, Emerson said that since "parsimony is not one of Nature's teaching and as efficiency is, it would be better to aim at efficiency first and leave parsimony to the generations to follow, who will be forced to make a virtue of necessity" (3–4). In Emerson's view, humans have learned prodigality, but not efficiency, from Mother Nature. Human nature, in other words, is wanton, wasteful, and short-sighted, only sometimes mitigated by over-correction through parsimony.

Human nature needed to learn efficiency for its wealth and salvation, and Emerson made it an engineering job to teach this lesson for Mother Nature:

> To attain the high efficiency of the atomic energy of the fish, the high mechanical efficiency of the bird, the high lighting efficiency of the firefly, is not an ethical or financial or social problem, but an engineering problem; and to the engineering profession, rather than to any other, must we look for salvation from our distinctly human ills, so grievously and pathetically great (5).

Because engineering practices were based on pure scientific knowledge, and because pure scientific knowledge had as its goal truth and man's domination of nature as decreed by God in the Book of Genesis, engineer Emerson could assert his authority over religious leaders, philosophers, economists, educators, politicians, etc. He could assert that engineers were the saviors of humankind—the holders of natural and divine knowledge that would be transmitted through technical writing. Emerson exemplified this conflation of science, nature, and religion in an example of the Prussian Army's reorganization for efficiency under General von Moltke in the mid-19th century.

Emerson described how von Moltke conquered Germany by using a modified military hierarchy that added a staff function to the traditional line organization:

> Germany succumbed helplessly before the genius of Napoleon in 1806, but less than two generations later von Moltke had remodeled the oldest of all organizations—military—by adding to line organization the principle of developed staff organization, and it is staff organization that has made Germany during the last 40 years easily the preëminent military power in the world (41–42).

Emerson later provided more details about how this staff was to function in relation to the line workers:

> The theory of a general staff is that each topic that may be of use to an army shall be studied to perfection by a separate specialist, and that the combined wisdom of these specialists shall emanate from a supreme staff. The specialist knows more about his one subject than all the rest of the army put together, but the whole army is to profit by his knowledge (60).

In this system, each staff member contributed to that "general fund" of knowledge that Rickard described in his first technical writing textbook, so that the whole organization could "profit by his knowledge." In this way, organizational efficiency could be maximized, opponents could be overcome, and individuals could prosper from the general increase in profits as they were distributed at factory owners' discretion. The key to enlarging the general fund of knowledge that would lead to maximum efficiency, however, was adding a staff function to the line organization and using technical writing to transmit staff members' specialized knowledge throughout the organization and throughout time. This, according to Emerson, was von Moltke's contribution to management science: "It is von Moltke's tremendous gift to the world that, although a soldier hampered by tradition, he applied to the army the other type of organization, the functional type, which ought always to have been used in business" (*Principles* 21).

Emerson asserted his belief that management systems based on line and staff functions were the key to maximum efficiency in complex modern organizations:

> Bismarck died humiliated; von Moltke is no more; but their business teachings live, and the modern German Empire whose every activity puts Great Britain into a senseless panic is next to the greatest example the world has ever seen of the result of modern business principles applied to the development of a modern world power.
>
> The greatest example of the power of rational organization and efficiency principles is not in the German upbuilding, but in the Japanese actual creation in a single generation of a great world power (*Principles* 19).

As noteworthy as the German and Japanese examples of world power building were, for Emerson they were cases of a universal law of nature: "To preserve the adult individual, Nature uses staff organization; to preserve the race, Nature uses line organization" (*Basis* 55). In Emerson's writing of history, von Moltke's army was victorious because it followed nature's laws of efficiency, which were set out by Emerson in twelve principles. Japan "put von Moltke's organization into effect . . . and also applied the twelve principles, which they had probably independently recognized" (*Principles* 20). The result of Japan's implementation of efficiency principles, according to Emerson, was that it was able to conquer vastly larger China and Russia in 35 years (*Principles* 20). The lesson of efficiency that Emerson taught for nature was clearly one of conquest and domination through management systems based on military principles. In Emerson's view, these principles applied to business because, using the examples of railroads, "A railroad has all the action of war, even to its disasters" ("Manufacture" 483). In the engineering view put forward by Emerson, successful business and military organizations alike were based on natural laws of efficiency.

Emerson's views on efficiency and organization evidently struck a resonant chord in other engineers and managers of his time. His first book was reviewed favorably in the *Railroad Age Gazette*, where a reviewer noted that Emerson was "a specialist in organizing railway mechanical departments so as to get the utmost service out of every dollar expended" (855). The topic of management and military organization was also developed within the pages of the *Engineering Magazine* during 1912 in a series of three articles by Professor Edw. D. Jones. In the first of these articles, Jones argued for the importance of articulating principles of "business administration" based on theories of evolution and a general fund of knowledge for the profit of all:

> Throughout history, the survival of the fittest, as between nations, has been fought out, in part, on the basis of the ability to use organized and co-operative methods of action. What a wealth of experience has been gained—and lost. . . .
>   And yet we seem to have accumulated but a small reserve stock of knowledge on this important subject (2).

Jones sought to contribute to the general fund of knowledge about business administration by looking to military history for amply recorded examples of the workings of large, complex organizations. Although military history was well recorded, business histories were considered proprietary information and were not recorded for reasons of competition among businesses.

Because Jones was translating military history into business administration, he had to first make a case for this translation. Like Emerson, he based his rationale on the success of the German military organization: "An analogy exists between the present needs of the American business executive, into whose hands in a generation a great increase in power has come, and the needs of the German army officers before the development of that splendid system which made Germany the leading military power of the world" (2). Jones used the same line of reasoning that Emerson had used before him in asserting that managers in the United States at the turn of the 20th century were working in organizations based primarily on the traditional line organization of military hierarchies. These modern businesses, however, could benefit from improving their organizations based on the line-and-staff model developed by the Prussian Army under von Moltke and his followers. By improving their organizations according to this functional military model, businesses could improve efficiency, increase wealth, and find salvation for humankind. Jones linked these business goals with military communication, scientific observation, and the coinage of metals: "A knowledge of the military manual of arms would be a suggestive preliminary to motion study, and an examination of the national mints would suggest methods of preventing waste" (6). By the time Jones wrote in 1912, technical communication was inextricably linked with metals, coinage, and science—and these were inextricably linked with military and industrial management systems. From its history as secret knowledge of occult mining practices, technical writing had become the lingua franca of science, and science had branched out into management engineering. Management engineering, in turn, looked to military history for models of efficient large-scale organization and technical writing provided the control mechanism for these organizations. Concurrently, technology was applied to technical writing, with the goal of improving the efficiency of technical writing and management systems.

NEW COMMUNICATION TECHNOLOGIES SUPPORT SYSTEMATIZED MANAGEMENT

Since management systems depended on technical writing as a control mechanism, system efficiency was directly influenced by the efficiency of technical writing. JoAnne Yates has argued that systematic management was able to spread from shop floors to business offices because of the communication technologies that developed concurrently with those systems: "New technologies contributed to the specialization of office skills and consequently created an opportunity for applications of scientific and systematic management to the office as well as to the factory floor" (64). More than spreading systematic management to another site in a large and complex organization, communication technologies enabled systems to function on the factory floor *because* systematized factory management was supported by systematized office management. Just as machines and workers became specialized and standardized to gain maximum efficiency from them, communication also became specialized and standardized through the implementation of office appliances and reporting forms. Among the office appliances that had great impact on the efficiency of technical communication within management systems in the late-19th and early-20th centuries were calculating machines, typewriters, copying methods, and filing systems. Along with these appliances, standard communica- ✳ tion forms were developed to decrease the time needed for recording, reporting, and comparing information.

By the 1890s, calculating machines were used in business settings, as well as in engineering and scientific practices. Business historian James Cortada claimed, "More than one dozen vendors [of calculating devices] operated during the 1890s" (43). These machines enabled improved efficiency in accounting departments, since they could manipulate numbers much faster than even the "lightning calculators," who Cortada described as "people who could add long, wide columns of numbers rapidly and even entertained with this skill" (27). Even "lightning calculators" had their speed limitations, however, which were overcome by the implementation of mechanized calculators. With faster and more efficient calculating, clerks and accountants within management systems could manipulate the increased volumes of information stemming from the need for more records to control the system. As Hamilton Church described in "The Meaning of Commercial Organisation"(1900), management system control required that all aspects of the organization be quantified, which in turn resulted in more data for analysis:

> With the growth of competition the necessity for co-ordination and of an accurate and swift presentation of results is more and more imperative. . . . Everything should be the subject of forecast as to financial results, and of prearrangements as to the actual carrying out. And when it is completed, the records of what did actually take place should be capable of comparison with what was intended to take place. Control then becomes a living reality (397).

Calculating machines made possible the manipulation of large amounts of operations data generated by attempts to control complex management systems through technical communication. Standardized forms also facilitated this data gathering and manipulating. One early form—the time ticket—was made necessary by pay rate systems, such as piece work, which relied on workers presenting records of their production in order for clerks to calculate their wages. An example of one such time ticket was included a paper entitled "The Premium Plan of Paying for Labor" that F. A. Halsey presented to the ASME in 1891. Halsey explained how this ticket was to be used: "This ticket is issued by the foreman, the blanks at the top being filled up by him. If desired as a check he punches a hole on the line, indicating the hour when the work is given out, repeating the same when the work and ticket are returned. The record of the time is kept by drawing a line between the various hour marks, an operation which the most illiterate can perform" (763). The foreman in this system was the forerunner of the mechanical time clock, performing the time-keeping function to control workers' pay. The time tickets thus generated could then be routed to clerks with calculating machines, where those records were efficiently transformed into wages for the workers and cost-of-labor records for managers to compare to other cost records. The time ticket provided a way to inscribe time that even an illiterate could write. By drawing a line on the ticket, a worker could turn this inscribed time into pay, thus rendering the technical writing on the time ticket as coinage representing its exchange value as real wages.

Typewriters also contributed to increased efficiency and standardization in record-keeping and communication. By 1890, the U.S. government reported that there were 30 firms manufacturing typewriters (Cortada 18). Although in 1888 businesses were buying 1,500 typewriters each month from Remington Standard Typewriting Company alone, typewritten business documents were not always well received. Cortada noted that "some people were insulted if they received a typed letter rather than a handwritten note and complained about the impersonal quality of the communication" (18). Another drawback to using typewriters in businesses was the lack of trained typists. The new typewriting technology, however, provided an opportunity for women to train as typists and enter the workforce without displacing men. The trend toward specialization combined with this new technology led to a new category of clerk whose functions were limited along the lines of Taylor's functional management model. Ellen Lupton described this new category of clerk:

> Through its newness as an object, the typewriter enabled managers to clear the table for the entrance of a new class of clerical workers with highly circumscribed roles to play. Whereas the traditional clerk often had been responsible for mentally *composing* as well as physically *writing* a text, workers in the mechanized office were assigned limited functions as stenographers (who captured an executive's spoken words in shorthand) and typists (who mechanically transcribed such words). . . . The new system . . . saved the high-cost time and effort of the managers, while a lower-paid crew of clerical workers generated a huge volume of legible, uniform documents (Lupton's italics, 44).

The legible and uniform documents that typists generated were gradually accepted by businesses and their customers. By the 1920s, according to Cortada, "[h]andwriting declined as typing spread, and with it, the personal distinctiveness that had actually grown with literacy during the nineteenth century was lost" (23). People became accustomed to less personal documents and the efficient production of these impersonal documents spurred further efficiencies. Cortada noted that the "size of typewriters contributed to the standardization of the size and shape of correspondence paper and envelopes" (23). Yates recounted how "the large workforce of trained typists and secretaries helped standardize the formats and conventions for the new genres of internal written communication" (44). In "Shop Management" Frederick Taylor described how typewriters could be used to standardize instructions for workers:

> The instruction card can be put to wide and varied use. It is to the art of management what the drawing is to engineering, and, like the latter, should vary in size and form according to the amount and variety of the information which it is to convey. In some cases it should consist of a pencil memorandum on a small piece of paper which will be sent directly to the man requiring the instructions, while in others it will be in the form of several pages of typewritten matter, properly varnished and mounted . . . so that it can be used time after time (180–81).

Typewriters allowed clerks to generate documents much faster than was possible by hand, at least doubling the rate of writing by hand. Typewriters also contributed to the specialization and standardization of communication functions, especially because they produced uniform documents of near-print quality. These standard, uniform documents were well suited for recording and conveying the detailed information upon which systematized management relied.

Taylor follower Daniel Vincent Casey described one effect of standardizing communication within a management system in "Putting Standards into Practice" (1911):

> Formerly, as in most shops, the mechanics did a large part of the planning how work was to be done. . . . Today, the workmen do no planning. Every detail of work on every job is thought out for them and put down in unmistakable black and white. . . . Each operation has been standardized: the standards are either carried in the planner's brain or in a convenient file: the instructions card carries these to the workman and his gang-foreman (67–68).

The instructions card was the agent carrying standardized operations information to the worker. Complementary reporting cards were used to record machine performance, workers' performance, and supplies used. Other reports would be used to record financial calculations regarding machine maintenance and wear, workers' wages and personal relationships, and cost of supplies. The management system relied on generating and storing forms and records in ever increasing numbers. This, in turn, required improved technologies for copying and filing documents.

JoAnne Yates traced the development of copying and filing documents from the time when outgoing letters were hand-copied onto blank pages in a bound copy book in chronological order. During the 19th century, this hand copying was supplemented by letterpress copying, in which outgoing letters were written in copying ink and copied onto dampened blank sheets of a bound copy book using pressure applied through a vice-like mechanism. Letterpress copying worked to standardize and specialize communication functions, as Yates described effects of press technology: "A letter press reduced the labor cost, both by decreasing copying time and by allowing an office boy to do the copying once performed by a more expensive clerk. At the same time, it eliminated the danger of miscopying" (28).

With the introduction of the typewriter, copying technology was to advance further through the development of carbon paper and mimeographic equipment. According to Cortada, a U. S. patent for carbon paper to use with typewriters was issued in 1872, approximately the same time as Remington manufactured the first typewriter (23). Shortly thereafter, A. B. Dick began to manufacture a mimeographic machine. The development of carbon paper enabled typists to produce a small number of copies as they typed their original documents. Copies of outgoing correspondence could thus be made along with the original document, overcoming the need for copy books and allowing for outgoing documents to be filed in alphabetical as well as chronological systems. The development of mimeographic equipment enabled clerks to make many copies of a document, which was especially useful for distributing documents internally. Yates noted that mass duplication of documents allowed managers to "distribute written rules, instructions, and announcements to large numbers of employees" (50). But how were large numbers of these internal communications to be stored?

In 1909, the Shaw-Walker Company introduced new filing equipment in which papers were stored vertically, as they are in most offices today, instead of horizontally as they were at the turn of the century. This vertical filing equipment enabled more papers to be filed, but offices needed a correspondingly modern filing system for clerks to retrieve papers efficiently. Yates noted that with the introduction of vertical filing equipment, a number of filing systems were suggested, such as numerical, alphabetical, geographical, and subject-based, but "[b]y 1938, one expert in business filing observed that alphabetical filing was clearly the most common system" (61). The use of standardized vertical filing systems, according to Yates, strengthened the control within management systems through improved internal communications:

> First, vertical filing systems organized by intended use rather than by origin and chronology allowed companies to create an accessible corporate memory to supplement or supersede individual memories. Accessible internal correspondence and records encouraged tighter administrative control over the growing companies. . . . Thus, vertical filing systems made internal communication a more effective tool of systematic management (62).

Efficient filing systems allowed the organization to replace individual memories for more standardized operations over time. Once the reports, time cards, and other records of an individual's performance were filed, these individuals could be inscribed for all time by the technical writing an organization had on file. Improved communication technologies had enabled organizations to control an individual's records well beyond that individual's life.

By the early 20th century, communication technologies had strengthened the control of large, complex organizations through management systems. These systems, modeled on scientific observation and Prussian military line-and-staff organization, were well suited to maximize operational efficiency through the implementation of production standards. They relied on constant examination, recording, and correction of worker and machine performance to maximize these operations. They also relied on the consent of individual workers to internalize standards and perform the constant examination in return for the promise of personal financial benefits. Because workers were treated as individuals, however, they gave up their identities as members of crews or other work groups in return for the promise of personal financial benefit to those individuals who successfully internalized the system's control.

From its origins on the shop floors, systematized management spread to the clerical office and even into the management ranks themselves. All workers—manual- and brain-workers alike—came within the systematic control of the organization. All workers—even the engineer/managers who designed management systems—became labor in the tensions between labor and capital. Workers relinquished the power to control their own work and became inscribed in a system controlled by technical communication.

TECHNICAL WRITING TEXTBOOK CODIFIES SYSTEMATIZED MANAGEMENT

By the beginning of World War I, office appliances were widely used in systematically managed organizations within the United States.

Managers routinely created the technical communications that controlled their organization's operations. Clerical workers performed the routine communication functions that maintained the stability of their management system. Yet the forms of recording and reporting developed for systematized management control were only codified at the end of World War II, when W. George Crouch and Robert L. Zetler wrote their textbook, *A Guide to Technical Writing*.

Crouch and Zetler taught technical students to be managers through their technical communications. The authors defended the study of technical writing explicitly by illustrating how communication skills could result in individual benefits for the students who would become engineer/managers within management systems. The authors also defended these management systems implicitly by illustrating how students were to use communications within these systems to control operations and maintain relationships between capital and manual, clerical, and managerial labor. Crouch and Zetler defended management systems and the technical communications that controlled them. But unlike the engineers who previously developed management systems and wrote about technical communication, Crouch and Zetler were professors of English who worked at the University of Pittsburgh. These authors did not have obvious expertise in management systems or technical communication, nor did they have the authority of coming from a discipline based on scientific knowledge. Instead, Crouch and Zetler claimed expertise in technical communication through their association with institutions traditionally thought of as having management system and technical communication expertise based on scientific knowledge: "the Schools of Engineering and Mines of the University of Pittsburgh, the Westinghouse Graduate Training Program, and the College of Engineering of the Carnegie Institute of Technology," as well as the "Director of Research Public Relations, United States Steel Corporation Subsidiaries . . . officials and employees of the Westinghouse Electric Corporation . . . Allis-Chalmers Manufacturing Co., Inc." (iv). By combining their expertise in English with other people's expertise-by-association in technical communication, Crouch and Zetler's alloy typified what would become a new model of technical writing textbook. After World War II, technical writing textbooks increasingly were written by authors with primary expertise in the field of English, with its liberal arts history. Because these authors sought to combine a traditionally nonscientific knowledge of English with science-based engineering knowledge, the technical writing textbooks they produced bore traces of historic tensions between these two types of knowledge and contemporary efforts to reconcile liberal arts with science.[2]

The authors began their *Guide* by presenting a typical day in the life of power engineer E. B. Nelson, who had been with Alpha Electric Corporation for at least five years. Crouch and Zetler described Nelson as working with a number of standard business communication forms: a letter from a field engineer regarding the temporary shutdown of a plant's rail line while a new steam line was installed; a letter from a customer asking Nelson to suggest equipment for a new mine; letters to customers dictated to Nelson's secretary; written telephone messages; interdepartmental and interoffice memoranda; departmental policies; carbon copies of order forms; materials forms for proposed designs; equipment catalogs; customer bulletins; employee suggestion forms. Nelson also worked with standard communications technologies: documents typed by secretaries; shorthand taken by a secretary in face-to-face dictation; dictaphone with recording cylinder for dictating to a secretary; carbon paper; file folders; telephone. The systematized office was established in Crouch and Zetler's introduction to communication for "technical students" (iii). But this textbook did not yet assume that the office systematized through technical communication was common sense to these students. The textbook, therefore, foregrounded instructions for how an engineer was to use communication as a management system control mechanism.

In providing narrative details about Nelson's use of the various forms of communication at the beginning of the textbook, Crouch and Zetler instructed students through Nelson's model of successful behavior for a corporate engineer as manager. The authors described, for example, how a manager should work with his secretary. They showed how Nelson had his secretary place a telephone call for him:

> He pressed the buzzer for his secretary.
> "Anne, will you get Mr. Ducey in Headquarters Engineering?"
> When Ducey's office answered and Anne was told that Mr. Ducey could speak to Nelson, she put him on the phone (5).

They showed how Nelson dictated a letter to Mr. Ducey with Anne in his office:

> He got his secretary to note an outline of his points on a scratch pad before he began to dictate his letter. . . .
> With these points as guides, he dictated the letter on the following page.
> Nelson read over the letter and signed it (5).

They showed how to use a dictaphone to record a letter for a secretary to type and how to give instructions to a secretary when leaving the office.

> He decided to dictate the letter on a dictaphone cylinder and then have his secretary type it while he was at the conference in the Powerhouse.
> He pulled the dictaphone to him and began his letter . . .
> "I'll just check over this bill of material on the drawing again . . . " Nelson told his secretary. "It'll take me about forty-five minutes. . . . If anyone calls for me, tell them I'll be back about 3:30. Don't tell them where I am. It won't take long if we're not interrupted" (12–13).

In these descriptions of Nelson working with his secretary, Crouch and Zetler showed an engineer's role as managerial agent for capital and a secretary's role as clerical labor. Nelson "gets" his secretary to make an outline for him; he "decides" how his secretary will do her work (i.e., in person or via the dictaphone); he "signs" a letter that is invisibly produced (by the secretary) and appears for Nelson's signature; he "tells" his secretary not to say where he has gone when he leaves the office, presumably putting his secretary in the position of lying for him if someone asked her where Nelson had gone. The relationship between Nelson and his secretary that Crouch and Zetler described is one in which Nelson managed his secretary, whose position existed to help Nelson generate the technical communications through which he helped to manage the work of Alpha Power Corporation. The relationship between Nelson and his secretary also reflected the relationship between Nelson and the technology that made communication more efficient. The secretary was like a dictaphone, except that she could receive a wider range of instruction than the dictaphone and she had reactions to her surroundings which must be managed to keep her good-natured. As an agent of capital, it was Nelson's job to manage the technologies producing technical communications in order to ensure efficient operations within organization and maximum profits for owners.

As a manager, Nelson needed to know at least as much about communications as his secretary in order to effectively manage her. This relationship was an important theme in the rationale for engineering students to study technical communication. A manager/dictator who cannot correct the mistakes of his secretary was not able to maintain the proper manager/worker relationship. He would not be able to use technical communication to control his clerical worker correctly:

> If a stenographer is so poorly trained that she makes obvious mistakes, the dictator should instruct her in correct methods. Every engineer should be able to do this. If he knows no more than his stenographer about form and content, he hardly deserves to have stenographic help (28).

But just as Nelson controlled his clerical workers through his knowledge of proper communication, Nelson's superiors controlled him through this same knowledge. If Nelson could produce correct communications, according to Crouch and Zetler, he would be rewarded, presumably through opportunities to move higher in the organization and earn higher wages. In one example, the authors describe how Nelson could be favorably noticed by a vice-president if he could correct the mechanics of a customer bulletin that was issued under the vice-president's name: "He revised the sentences which tended to run together; he made the antecedents of the pronouns clear; he corrected the spelling. After all, that letter came from a vice-president, and vice-presidents often have much to do with promotion" (17). On the other side of the coin, Crouch and Zetler described how as a young engineer, Nelson's poor letter writing got him into trouble with a higher-up:

> But after he had landed his job with Alpha Electric, he realized for the first time that the relationship between the shop and the main office was kept up largely through letters and memoranda. Some of his work became known by the executives of the company only through the reports he wrote.
>   Nelson remembered a comment E. W. Harrison, head of Power Engineering, offered one day. "Where the hell did you go to college, anyway? You write business letters like an eighth grader." That comment hurt (19).

Here Crouch and Zetler showed Nelson being managed by Harrison through the mechanism of correct business letter production. They went on to describe how, as a result of this incident, Nelson "had determined to master the art of putting ideas on paper" (19). In other words, Nelson had internalized the management system's emphasis on proper technical communication as a result of Harrison's management. This passage also illustrated Crouch and Zetler's recognition that communications controlled the relationship between management/capital and labor: "[T]he relationship between the shop and the main office was kept up largely through letters and memoranda." Within this short passage the authors show Nelson as a manager working within a management system controlled by technical communication—both controlling shop workers through the letters and memoranda he generated and being controlled through the application of communication standards from the managers above him. Nelson was the ideal engineer/manager who had internalized the management system—and whom the students reading Crouch and Zetler's textbook were to become.

Crouch and Zetler further reinforced the idea of engineers as managerial workers, echoing Frederick Taylor's and other management systematizers' emphasis on the worker as an individual. The systematizers were managed by the systems they created. Engineer/managers were separated from capital/factory owners, the better to control the managers themselves: "He is first an individual and after that an engineer, and in this day of big shops, human contacts are more likely to be through letters or telephone calls than by actual person-to-person talks" (16). As a model engineer/manager, Nelson had internalized his position as an individual within the management system. This led Nelson to seek personal reward from the department heads and vice-presidents in his corporation. Instead of seeking to contribute to a general fund of knowledge—a rationale for studying technical writing that Rickard put forward in 1908—Nelson was controlled by the promise of individual rewards. The earlier notion of communal scientific knowledge leading to the betterment of general living conditions for humankind had been superseded by the notion of an individual's scientific knowledge leading to the betterment of living conditions for that worker and his family.

Within the economy of the management system, a worker's knowledge could be represented by a dollar value. In the example previously cited of Nelson's improving a vice-president's bulletin, Nelson's technical writing knowledge could result in a promotion and an increase in pay. Similarly, Harrison's chiding Nelson for his poor letter-writing could result in Nelson losing his job and his pay, or being passed over for promotion and losing a raise in pay. Workers' knowledge, too, could have a dollar value, although its value was determined by managers and would probably be valued less than a manager's knowledge. For example, Crouch and Zetler described how Nelson talked with a "group leader" who "had no technical training" (13). This group leader, Henry, suggested an improvement to an Alpha Electric installation and Nelson responded in this way:

> "That's a good idea, Henry, and it'll save some money and quite a bit
> of work. . . . [Y]ou write your idea on a suggestion blank; I'll accept it
> and the company ought to give you five dollars for it."

On the way back to the office, Nelson was thankful he'd never adopted the opinion that was so general among clerical workers: that whatever a shop man thought, was wrong. . . . When Nelson had first come to Alpha Electric, he had had much the same idea. What could a man without technical training or experience tell him? But his first year's work, when he had been on night turn on the B-7 test floor, showed him that there were quite as good brains among the shop people as there were in the office; they simply followed a different line (13–14).

While Nelson's good ideas, if noticed by a vice-president, could result in a promotion and raise, Henry's good ideas were worth five dollars—if the company agreed to pay him extra for his ideas. Although there may be good brains among the manual-workers, these workers would always follow "a different line" from the brain-workers. The word "line" here implies a Darwinian separation of manual- and brain-workers within the management system as a natural consequence of evolutionary development. Manual-workers were more fit by nature for the heavy labor they did while brain-workers were more fit for sedentary labor. Thus, their positions within the organization were decreed by nature and should not be disputed. Manual-workers were placed on a different line in an organizational chart and would only intersect a brain-worker's line at specific points. It was at these points that brain-workers needed to manage manual-workers. Part of this management was prospecting manual-workers' lines of practical knowledge for possible knowledge that could be valuable to the manager and to the organization.

Crouch and Zetler articulated the moral to Nelson's example in his position of manager as diviner, prospecting workers' brains for valuable ore that could be refined into scientific currency through technical communication: "Nelson's meeting with the wiremen . . . is important because it implies a frame of mind that every young engineer should have. It rarely costs much in time to listen to suggestions. Perhaps nine out of ten suggestions from nontechnical men are useless, but that tenth may be of great value" (17–18). This prospecting frame of mind maintained the relationship of managers controlling workers as natural resources within the management system, while being on the lookout for that rare specimen of workers' knowledge that could increase the efficiency of the organization. Presumably, the manager who mined this bit of knowledge and transformed it into something valuable to the organization would also benefit individually through possible promotion and increased pay. The worker also might benefit individually if managers decided to pay him for his "suggestion." The worker's benefits would probably be less than those accorded to the manager, however, because in a management system the manager's knowledge is more powerful than that of the workers. It is the manager's knowledge that is translated into the technical communications that control the workings of the system.

7

Technical Writers Mint Counterfeit Scientific Knowledge: Strained Relations between Technical Writers and Engineers

New frontiers of the mind are before us, and if they are pioneered with the same vision, boldness, and drive with which we have waged [World War II] we can create a fuller and more fruitful employment and a fuller and more fruitful life.
　　　　—FRANKLIN ROOSEVELT, *Science, the Endless Frontier*, 1945

Industrialism . . . diminishes the freedom of the individual in relation to the community, but increases the freedom of the community in relation to Nature. That is to say, the actions of the individual, at any rate, in the economic part of his life, become increasingly controlled by the actions of the community, or by some large organization such as a trust; but the actions of the community become less and less controlled by the primitive necessity of keeping alive. Hence individual passions, such as those which produce art and romance, tend to die out, while collective passions, such as those which produce war, sanitation, and elementary education, are liberated and strengthened.
　　　　—BERTRAND RUSSELL, "Where Is Industrialism Going," 1928

During the late 19th century, technical writing was the medium for engineers and scientists to convey their knowledge to other professionals—to contribute to Rickard's "general fund" of scientific knowledge. Technical writing, therefore, was the currency that represented value inherent in scientific ideas. Other branches of writing, like English composition and literature, were separate from technical writing and were properly grouped with other disciplines, such as history and philosophy, which depended on humanistic bases of knowledge. Technical writing, on the other hand, bore the stamp of science and was considered to be a part of scientific practice as long as engineers and scientists did their own writing.

When engineer-designed industrial management systems began to be studied as a specialized field during the early 20th century, business communication separated from both English composition and technical communication to become a distinct academic concentration. With this separation, managers became increasingly distinct from engineers and clerical workers took over much of the job of generating the communications through which the functional-military management system was controlled. These clerical workers, however, had a constrained role within the management system and did not use as much individual discretion in generating communications as clerks had in pre-systematized business organizations. Modern clerical workers served as transcribers of managers' words or generators of routine formulaic communications. Because of their constrained and routine functions, clerical workers could be paid at lower rates than managers, thus freeing managers to spend their higher-paid time doing more important tasks than typing and filing routine communications within the management system. Because clerical workers freed managers to spend their time on analyzing and planning operations for increased efficiency and productivity, creating a new occupation of lower-paid clerical workers proved profitable even though this class of worker grew at a rapid rate during the first half of the 20th century.

Like business communication, technical communication become increasingly separate from engineering and was constructed as a specialized field of English studies. After World War II, a growing cadre of technical writers were not engineers, but were writers trained in the specialty of technical communication.[1] This specialization was due in large part to the boom in technology during and after the war, which pressured engineers to develop technology instead of communicating about it. Engineers were considered to be the high-priced workers who were better used in developing the technology that would improve general living conditions and stabilize democracy. Lower-priced writers could take care of communicating these technical developments. In a trend reminiscent of the management/clerical worker separation, engineering functions were split from communication functions in hopes of greater efficiency and productivity for technological development.

This specialization of technical writing as a field of study within traditionally humanistic academic departments led to tensions concerning who ought to produce engineering knowledge—engineers whose knowledge was based on pure science or writers whose knowledge was based on liberal arts. These tensions were articulated through discussions about the role of the humanities within technical curricula—a 20th-century iteration of the Huxley/Arnold debates in the previous century. Unlike the 19th-century debate, however, this recent discussion positioned technical education as the norm against which the humanities must defend itself. In this positioning, science is analogous to the god-king that, in "Plato's Pharmacy," Jacques Derrida argued was the source of the pure unwritten Word:

> God the king does not know how to write, but that ignorance or in-capacity only testifies to his sovereign independence. He has no need to write. He speaks, he says, he dictates, and his word suffices. Whether a scribe from his secretarial staff then adds the supplement of a transcription or not, that consignment is always in essence secondary (76).

Derrida further argued that although the god-king claims sufficiency for his unwritten word, as a supplement, writing reminds the god-king of the insufficiency of his word that needs to be written by another. Writing as a supplement is thus the knowledge that destabilizes the god-king's word. Derrida likened this relationship between the god-king's word and writing to the relationship between Thoth (Hermes) and Ra (the sun god): "The god of writing [Thoth] thus supplies the place of Ra, supplementing him and supplanting him in his absence and essential disappearance" (89). But in supplementing/supplanting the god-king's word, writing can change the word and deceive the reader. Derrida quoted this passage from Plato's *Republic* in which writing addresses the god-king, promising to deliver the god-king's "offspring" or "interest" on a debt:

> Well, speak on, he said, for you will duly pay me the tale of the parent another time—I could wish, I said, that I were able to make and you to receive the payment, and not merely as now the interest. But at any rate receive this interest and the offspring of the good. Have a care, however, lest I deceive you unintentionally with a false reckoning of the interest (83).

Because writing can deceive—intentionally or unintentionally—it is the destabilizing supplement to truth that cannot convey truth.

Citing Plato's *Protagoras* and *Philebus*, Derrida argued that "writing is essentially bad, external to memory, producing not of science but of belief, not of truth but of appearances. The *pharmakon* [of writing] produces a play of appearances which enable it to pass for truth, etc." (Derrida's italics, 103). Because "writing is essentially bad" and deceptive, it must be banished outside the boundaries of sacred knowledge to purify that knowledge, just as scapegoats were banished from ancient Greek communities to purify the righteous inhabitants in times of crisis:

The restoration of internal purity must thus reconstitute . . . that to which the *pharmakon* should not have had to be added and attached like a *literal parasite* . . . In order to cure the [word] of the *pharmakon* and rid it of the parasite, it is thus necessary to put the outside back in its place. To keep the outside out. This is the inaugural gesture of "logic" itself . . . Writing must thus return to being what is *should never have ceased to be*: an accessory, an accident, an excess (Derrida's italics, 128).

Derrida's description of how dominant knowledge works to purify itself of writing explains how science worked to purify itself of art and the humanities in the 20th-century iteration of the technical education/classical liberal arts education debate. This recent version of the debate featured science as the sacred/dominant word that risked degradation from the art of writing. Science, thus, sought to purify itself from this intrusive parasite of art by assimilating only the practical aspects of art that could be compatible with science and eliminating any incompatible impracticalities.

Because science could pressure art to be practical and mass producible, art became detached from its historical roots as a religious or cult fetish. Art relied on science, not religion, for its meaning. Walter Benjamin described the outcome of such a modern transformation of art: "When the age of mechanical reproduction separated art from its basis in cult, the semblance of its autonomy disappeared forever" ("The Work" 226). And further,

[F]or the first time in world history, mechanical reproduction emancipates the work of art from its parasitical dependence on ritual. To an ever greater degree the work of art reproduced becomes the work of art designated for reproducibility. . . . But the instant the criterion of authenticity ceases to be applicable to artistic production, the total function of art is reversed. Instead of being based on ritual, it begins to be based on another practice—politics ("The Work" 224).

When art in the industrialized 20th-century United States became defined by scientific notions of practicality, art was connected to religion only through science. And they were both connected to the politics of systematized management. The only way for arts or humanities to gain legitimation in such a culture was to emulate the sciences and find what was practical in the arts.

English composition found its practical nature late in the 19th century through its use as a tool for upward social mobility. By training middle-class workers to use correct English, composition instruction could prepare engineers, for example, to be favorably noticed by higher-ups and become successful managers themselves. Those engineers who succeeded in climbing the social ladder could then judge the English of newcomers who might aspire to join them inside the upper levels of society. Other humanistic pursuits, such as literature and history, could supplement an engineer's technical education to give him an understanding of human relations that would also prepare him to become a manager, move up in the functional-military management system, and reap personal financial benefits. By emphasizing their practical applications within business, English composition and other traditionally humanistic fields could find a supplementary place within technical curricula.

As Derrida argued, however, dominant knowledge acknowledges its insufficiency through the need of a supplement. Scientific knowledge and technical education, therefore, admitted their vulnerability while arguing for the place of practical humanistic education within their curricula. Similarly, management acknowledged its Achilles heel of communications through its need for a new class of clerical workers to transcribe managers' words and maintain routine management-system communications. Engineering likewise acknowledged the vulnerability of its scientific currency—technical communications—through its increasing dependence on specialized writers trained in the classical liberal arts tradition. These specialized writers highlighted the fragile nature of human-made technical and scientific knowledge. They also rendered scientists and engineers mute by supplanting them as the communicative agents in their fields. But perhaps most troubling, specialized technical writers could replace the stamp of science with the stamp of speculation, thus devaluing the currency of scientific knowledge in the knowledge/power system. One reaction to this threat of instability was an attempt to return to the time a century earlier when engineers did their own writing and systematized management promised unlimited progress and social improvement.

OFFICE MANAGEMENT
BECOMES A SPECIALIZED FIELD

The trend toward systematized management, which began in the last decades of the 19th century, blossomed into what American Management Association Editor-in-Chief W. J. Donald called "the management movement" in his preface to the *Handbook of Business Administration*

(1931): "It is our hope, too, that this 'Handbook of Business Administration' will find its way into the hands of many business executives who have heretofore not been associated with the management movement either in America or abroad" (vi). By the 1930s, systematized management was mature enough for Donald to compile a "digest of the best of [the American Management Association's] publications . . . also contributions covering the field of modern management methods" (v). By Donald's own admission, though, not all business executives had "been associated with the management movement." Systematized management was not clearly the dominant form of organization by the 1930s, but it was on the verge of dominating business practices. With this dominance of management systems came the mechanism of technical writing that both communicated knowledge about management systems and enabled these systems to control workers and their work. The skeleton of systematized management can be seen in the contents of the *Handbook of Business Administration*, which included (in order of presentation in the handbook) sections on marketing, financial management, production management, office management, personnel management, and general management. The field of systematized management science was taking shape and the American Management Association's handbook was designed as an autodidactic tool for standardizing, measuring, and evaluating specialized management practices.

The office management section of the *Handbook of Business Administration* included an article by M. B. Folsom, assistant treasurer of the Eastman Kodak Company. This article, "The Field of Office Management," gave an overview of the areas of concern falling within this management specialization. The office in Folsom's model was the hub of management system communications, housing the office appliances, forms, internal and external correspondence, files, centralized mailing, and clerical workers to generate and maintain the communications that controlled the organization's operation. These clerical workers, in turn, were controlled by other clerical workers and managers enforcing standard methods, production standards for clerical workers' output, and standardized salary limits. Each clerical worker's production was compared to these standards by workers in personnel, finance, production, marketing, etc. to evaluate the individual's performance and determine pay and promotion possibilities. Clerical workers generated the technical communications through which the system was controlled; they themselves, in turn, were controlled by these technical communications. This panoptic organization relied on the constant examination of workers through reporting systems that tracked the minute details of their performance. These workers internalized the system and participated in this constant examination, reporting on their own performance and that of other workers. The system relied on workers viewing themselves as individuals whose performance was examined and corrected separately from that of other workers—and who would be rewarded or punished in isolation.

In the first half of the 20th century, the scope of office work and the number of office workers increased significantly. Elmer Grillo and Charles Berg quoted an often-cited statistic in their introductory chapter to *Work Measurement in the Office* (1959): "Almost every article on the subject of office costs in recent years has stressed that while in 1900 there were something like 10 office employees for every 100 workers in the factory, today the figure has grown to about 25 for every 100" (4). Grillo and Berg attributed this growth in office workers to mass production, more sophisticated management demanding "a steady flow of statistics and other data undreamed of years ago" (4), the growth of accounting, increased government reporting requirements, increased emphasis on public relations, and "a tremendous growth in industries which employ a great deal of office help" (5).[2] These authors emphasized this growth in clerical labor to highlight their argument that offices should be scientifically managed along Taylor's model of shop management: "Office-management literature has been giving more prominence to the use of procedures studies to eliminate unnecessary work and to install improved work flows. But the second area, work standards for the office, is in need of greater attention" (5). Grillo and Berg's generalizations about office-management literature indicate that office management was a specialized field of management by the late 1950s. Offices had grown and changed since the turn of the century, yet Taylor's scientific management continued to be a widely used model for efficiency in production, whether on the shop floor or in the office.[3]

The field of office management was concerned with how to control clerical operations and workers for maximum efficiency, quality, and productivity. This goal was complicated, however, by the volume of technical communications that were produced by organizations using systematized management. Elmer Grillo summarized the paperwork situation in *Control Techniques for Office Efficiency* (1963):

> There is one thing which characterizes our modern business forms today, regardless of size, and that is is [sic] the volume of paper work necessary to run them. This flood of paper and detailed records seems to be peculiar to our own day. . . . While written records are

undeniably essential to control our modern business firms, an increasing number of organizations are wondering if paper work has gone too far (1).

Grillo's solution to the paperwork problem was to offer a variety of office management techniques and advise managers "actually to install and use a wide selection of techniques covering both routine office work and nonstandard office activities; both clerical work and supervisory, managerial, and professional activities" (viii). As in Grillo's earlier co-authored book on office management, in *Control Techniques* he advocated a range of scientific management methods for work measurement, standards implementation, and cost-value analysis. In this later work, however, Grillo illustrated how scientific management systems could be applied to management and professional workers, as well as clerical and shop workers. By the 1960s, the implicit control of all workers that was designed into management systems was taken to be a natural state of the systematized worker. Individual workers in the factory and office alike had learned to internalize the system in return for the benefits they could gain for themselves and their families.

Like the shop workers in the early 1900s who chafed under the new constraints of early management systems, however, the clerical and "brain-workers" (as Taylor called them) found the process of internalizing systematized management control to be an unpleasant experience. Grillo described the dissatisfaction of clerical workers after World War II:

> Perhaps the shortage of office help since World War II, coupled with an awareness of the increasing restiveness of the white collar worker in viewing his diminished pay and prerogatives compared with the factory employee, has given impetus to this preoccupation with human relations. It has been an effort to create a happier, more congenial, and therefore more productive working atmosphere (3).

The engineers who discussed Henry Towne's gain-sharing pay plan at the American Society of Mechanical Engineers meeting in 1888 were concerned with the question "How can we at once get the workman to squeeze a little more out of himself and at the same time be good-natured in doing it?" (616). It seems that the same question was being asked about the clerical workers after World War II, except now the goal was to create a "happier, more congenial . . . atmosphere," with the inference that this atmosphere would lead to happier and more productive clerical, management, and professional workers.

This concern with human relations was sidetracked, though, by a parallel development in office management technology: automation. This trend to automation of management communications was noted by Niles, Niles, and Stephens in *The Office Supervisor* (1959): "There are more than fifty companies manufacturing and selling electronic computers in the United States alone. Their application to office work in business and government grows daily" (11). These authors' focus was on training supervisors to have good human relations skills while also maintaining production standards:

> The supervisor's job is more than giving out the work and following up to see that it is done. He must make each task assigned worthwhile, encourage workers' ideas, and help make their work a satisfying experience. He must also meet production goals. He cannot be effective if he does not show genuine interest in his subordinates (7–8).

In this respect, Niles, Niles, and Stephens followed in the familiar line of management reasoning wherein managers must make workers good-natured about working at maximum production. With the addition of automation to this situation, however, human relations took a back seat to increased efficiency for the management system through new technology:

> we wish to emphasize how [automation] affects the office supervisor and those persons with whom he must deal. . . .
> 5. *More skill is required, even on lowly clerical jobs. Many jobs of routine, repetitive work are abolished* in an office with electronic data processing. . . .
> 8. *Those* [clerical workers] *who are not upgraded suffer to a greater or less extent.* Some must be downgraded or dismissed either because fewer jobs remain with the new systems or they lack the capacity for the change-over. . . .
> 10. *The general concern with systems and with machines creates a need for the alert supervisor to look for new and better ways of doing work* and to cooperate understandingly and willingly with those in the company who specialize in planning and systems work (Niles' italics, 12–13).

Along with factory and clerical workers who must be good-natured about internalizing systematized management control, managers now must also be good-natured in the face of increased control through electronic data processing technology. The supervisor, in addition, must have a "genuine interest" in the clerical employees who would be downgraded or dismissed in compliance with the demands of increased technological control of office management, while still complying "willingly" to the systems planners who gained control through technological communication. The electronic age was coming to office communications and management systems.

In his 1970 preface to *Office Management and Control*, George Terry began, "In the field traditionally known as office management, change and evolution continue to make this area one of the most dynamic in our economy" (vii). So much change was taking place in the field of office management, Terry distinguished between the automated practices of his day and the non-automated practices "traditionally" associated with management systems control. He went on to reiterate the time-honored view of the relationship between management and technical communications, this time with a new twist:

> Information is a basic and universal need of a manager; it is an integral component of everything he does . . . The computer and other office machines are revolutionizing our approaches to information handling. Likewise, innovations in the use of systems and procedures, automated equipment, microform [sic], office work measuring, quality controlling, and employee motivating are sharpening our focus and efficiency in meeting these information requirements (vii–viii).

In Terry's view of the relationship between management and communication, making workers good-natured—"employee motivating"—was a step toward acquiring necessary information for managers and the computers with which they worked to generate communications. This automated version of office management placed the goal of communication in the place of the traditional systematized management goal of productivity. Workers were to be efficient information gatherers so that managers could have the necessary information to feed to the computer, which would ensure that the system ran at maximum efficiency. The traditional management goal of productivity may have been assumed as industrial common sense by 1970. But in changing the focus of office management from productivity to communication, Terry highlighted the central role of communication in controlling the management system, and put forward the promise of automation to make communication a more efficient control mechanism.

As office management became a separate field of study during the first half of the 20th century, so did business communication become a separate field from either composition or technical communication. As

early as 1917, George Hotchkiss and Edward Kilduff asserted this separation in stating at the beginning of their *Handbook of Business English,*
"Business English includes all written messages that are used in trans-
acting business. Because of its distinctive purposes and method it is
justly regarded as a separate branch of English Composition" (1). Kilduff is listed on the *Handbook's* title page as being a professor of business English at New York University, thus indicating that the university
also recognized this as a separate field of study.[4] But in their need to assert the separation of business English from other writing instruction,
Hotchkiss and Kilduff also suggested that this separation was not yet
within the common sense of either businesspersons or writing instructors. Another indication of the newness of this field can be found in the
handbook format that these authors intended as an autodidactic tool.
Just as Donald's *Handbook of Business Administration* had been an early
compilation of all the important information on this newly developing
field of study intended as an educational tool for managers, so was the
Handbook of Business English intended for use by businesspersons as
well as students: "Both in the office and in the classroom [the *Handbook]*
has proved a time-saver in correcting the faults of correspondents and
students and in guiding them toward a better mastery of the art of writing" (x). That these two early handbooks were intended for use by people already practicing management or business English suggests that
formal schooling in these fields had not been available for a long
enough time to be taken for granted in the practice of management.

By 1949, however, business communication was becoming a more
stable field of academic study. An example of this stability can be
found in the textbook *Business Writing,* which author Charles Carey
called a handbook: "This book is designed primarily for the world of
commerce. . . . Yet it is only a handbook, adequate in principle but de-
void of lengthy discussion. In every instance, ample opportunity is
given the student to put the theory into practice by means of appropriate exercises" (ix). Carey said he wanted his book to be "a desk aid for
handy reference" (ix), indicating that he expected people to use the
book in business settings. But he included exercises at the ends of his
chapters, which is more appropriate for classroom instruction than
business use. Carey also refered to the pedagogical benefits from examples of business communication that he included in the book:
"Good classroom procedure demands that you glance occasionally at
the actual business world to discover what men are saying, and what
reactions follow. For this reason many examples are utilized to illustrate the adaptation of theory to business practice" (ix–x). Here Carey
addressed students as the "you" in the classroom—readers not yet in

the "actual business world"—indicating that *Business Writing* was more a textbook in a recognized discipline than an autodidactic tool. The subject position Carey thus provided for students instructed students to follow capital-producing practices in "the actual business world" and model their communications after those practices. As Lester Faigley has argued in *Fragments of Rationality*, student subject positions that writing instructors traditionally create in classrooms work to maintain existing relations of knowledge and power: "[R]elations of power are sustained by controlling the subject positions that readers are invited to occupy" (99). This subject position that Carey created allowed students to reproduce existing business practices as ideals without encouraging them to question those practices, thereby stabilizing existing cultural relations between labor and capital, workers and private property, wages and profits.

Carey's indebtedness to Hotchkiss and Kilduff's earlier work in defining the discipline can be seen in Carey's decision to include this epigraphic quote from Hotchkiss in his introductory chapter: "Business English isn't a language; it is a point of view" (xi). Thirty-five years after Hotchkiss first asserted the separation of business communication from English composition, Carey continued to argue for business writing as a specialized field of study. Carey further evoked systematized management's historical roots in the line-and-staff organization developed in the 19th-century Prussian Army when he began his preface by stating, "Language is a weapon which the business world uses to the hilt" (ix). Carey evoked management's history in mechanical engineering when he likened business communication to the steam engines with which his forebears had been so concerned: "To the progressive businessman, the English language is essential equipment, and a source of power on the highly competitive American scene" (ix). Carey also stated that the "terminology [of business writing] is merely a part of the routine machinery of commerce" (xi). The need for this "routine machinery of commerce," according to Carey, was brought about by much the same circumstances as Daniel McCallum had described in the growth of long-line railroads in the mid-19th century. Instead of railroad supervisors losing touch with their workers, in Carey's world of large businesses shopkeepers had lost touch with their customers:

> Today, business has widened beyond the confines of neighborhood markets. In the old days business was transacted only among neighbors . . . It was a matter of personal transaction. Faulty local expressions and other personal shortcomings were either overlooked or ignored. But those conditions no longer prevail. The manufacturer no

longer is acquainted with the customer. His market is now nation-
wide, even international. Dialects and local customs are not adequate
to current situations (xiv–xv).

Because businesspersons had lost personal touch with their customers,
business communication—the engine of American business competition—had to be more standardized and systematized than was necessary when people conducted business face-to-face. In the model of
systematized management, business communication became a specialized field of study and practice in order to standardize this mechanism
for business control. Business communication became a technology for
removing idiosyncrasies from the documents and verbal communications on which business and management systems relied. Like the
typewriters, carbon paper, duplicators, and other office appliances that
had been introduced over a half-century earlier, business communication worked to shape business transactions within standardized limits
of individuality.

After World War II, business communication was a specialized
field of study, but its roots were in the same military, engineering, and
railroad history from which technical communication was developed.
It also shared with technical communication a concern for practical
outcomes. As Cecil Williams and E. Glenn Griffin phrased this practicality in *Effective Business Communication* (1966), "Business communication differs from general communication chiefly in being closer to the
payoff—in being aimed more specifically at making something hap-
pen" (4). These authors saw business communication as being closely
tied to advertising and promotion, with payoffs for business and individuals alike. In giving students a reason to study business communication, Williams and Griffin echoed the argument put forward for so
long in technical writing textbooks, i.e., the study of business/techni-
cal writing will bring the student/worker personal financial benefits.
In their iteration of the argument, Williams and Griffin framed the rationale in this way: "The practicality of your studying communication
will soon become evident to you. Your first application and interview
for employment will be the proving ground. And, in subsequent employment, proficient application of what you have learned will bring
rewards, both personal and monetary" (4). Nelda Lawrence and Elizabeth Tebeaux also included this familiar argument in their 1982 work-
book *Writing Communications in Business and Industry*: "You should
never forget that your success in any organization depends as much or
more on your communication expertise as on your technical knowledge" (1). Linda Driskill began her 1992 textbook *Business and Managerial*

Communication with the same argument, which by now has become just another bit of common sense within the larger common sense of systematized management:

> Your success in business depends on how well you communicate. Researchers, managers, and students who work all agree that your ability to write, speak, participate in meetings, interact with people from other cultures and countries, and use computers and other media will affect whether you are hired and promoted. Businesses will no longer tolerate employees who are not productive (3).

By the 1990s, readers have accepted the connection between management control and communication that was once a concept in tension between individual and group identity, labor and capital, worker or management control of how work was accomplished. Driskill need only assert the relationship among communication, individual productivity, and success to readers who have already internalized the management system, even before entering the workplace. But why does this apology for business writing need to be made in 1992 in the same way that the apology for technical writing has been made since the first textbook was published at the turn of the century? A closer look at the development of technical writing as a discipline can illuminate this question further.

THE PRACTICALITY OF ENGINEERING IN TENSION WITH THE ENJOYMENT OF LITERATURE

In the early years of the 20th century, technical writing was one of the jobs carried out by engineers and scientists who needed to communicate knowledge to other people. As Rickard described this process in 1910, engineers and scientists had an obligation to contribute to a "general fund" of scientific and practical knowledge to further their work and the progress of humankind. They made their contributions directly to the fund through the currency of technical communications. In this scheme of knowledge-making, technical communication was within the fields of science and engineering and clearly relied on scientific knowledge for its stamp as genuine currency.

By the end of World War II, however, technical communication did not clearly fall within the realm of engineering and science as it had in earlier days. Textbook authors George Crouch and Robert Zetler, for example, were not engineers as Rickard had been; they taught in the

English Department at the University of Pittsburgh. Their 1948 *A Guide to Technical Writing* was organized more like a traditional composition text based on forms of writing rather than the topical organization of Rickard's earlier *Guide*. In the preface to his second edition, Rickard described his role as a former engineer who had become a better-than-average technical writer through his practical experience as a writer and editor. He wanted to contribute his knowledge of technical writing to the general fund so that other engineers could benefit from his experience: "I write as a scribe, without authority, except in so far as the members of my old profession will concede it to me from the nature of my present occupation [as editor of *The Mining Magazine*]; I speak as a student, not a master; as an amateur who has become a professional, but not a professor" (4). Because he was writing from his personal experience, Rickard organized the material in his book according to engineering concerns with language at the turn of the century: abbreviations, numbers, education, hyphens, word usage, titles, pronouns, journalistic language, grammar, etc. Readers of Rickard's *Guide* followed the author's path through the culturally shaped observations and interests of one engineer-turned-writer who presented his writing rules from the editor's desk, not the professor's lectern.

Crouch and Zetler, in comparison, placed themselves primarily within the academy by alluding in their preface to their experience as teachers, while also differentiating technical writing from general composition courses:

> The authors are aware that many technical students feel that the usual English composition course lacks direct application to their future professional life. . . . the materials for this book have been winnowed from work in English and Speech carried out by the authors, not only with students in technical schools, but also with technical men in industry (iii).

Working from this academic base, Crouch and Zetler organized their material in chapters according to formal categories: "The Business Letter: Form and Substance," "Types of Technical Letters," "The Technical and Semitechnical Article," "The Technical Report." To this they added "Speaking Techniques," a medium of technical communication reminiscent of the oral traditions of early rhetoric-based education; a chapter on "Language Usage" that included topics such as outlines, sentences, paragraphs, coherence, unity, and emphasis; and an "Index to English Usage" that included grammatical rules. This organization resembled earlier composition textbooks dating back to the late 19th century, which will be discussed below. But before moving to these earlier texts, it is noteworthy that Crouch and Zetler also included an appendix in which they presented a reading list of literary works for engineering students to read. They included this list in response to a notion that engineers were too specialized in their education and interests, to the detriment of the general human condition that they were to improve:

> The engineer is often criticized because he is not as widely read as the majority of professional men. To a large extent, his specialized education is responsible for this; while in school, his time has been taken up almost exclusively with scientific matters. After entering industry, however, he has as much free time as the average professional man, and in order to be generally well educated, he should read material in fields other than his own (393).

The author's three-page reading list included biographies, English and American novels, "Detective, Horror, and Scientific Fiction" (394), English and American plays, short stories, essays, and poetry. It was intended to supplement an engineer's scientific education, which was thought by many engineers, managers, and academicians to be deficient for producing the type of well-rounded engineer who could understand human issues.

In asserting that a technical education was narrow in scope, Crouch and Zetler contributed to an ongoing discussion of the shortcomings of a scientific education—a discussion situated in the continuing tension between the classical liberal arts curriculum advocated by Matthew Arnold in the late-19th century and the technical curriculum advocated by Thomas Huxley. It seems that Huxley's position had gained ground in this contest, as suggested by the fact that the 20th-century iteration of this discussion centered on scientific education. Instead of taking the classical liberal arts curriculum as the norm against which to position the scientific curriculum, the 20th-century argument implied that the scientific curriculum was the norm against which the liberal arts curriculum was positioned. Philip McDonald rehearsed this contemporary version of the liberal arts/science tension in his technical writing textbook *English and Science* (1929):

> Too rigidly scientific a point of view defeats the aim of life and and alienates the sympathies of friends and associates. A man who tries to be unduly practical is likely to end up by not being practical at all. The student who studies only those subjects for which he can foresee a useful application after graduation will come out of college badly

educated and unfitted for leading a full, normal life, with all that such
a defect implies in the missing of opportunity and the lack of support
from associates (156).

McDonald addressed his textbook to "engineers, scientists, and most
business and professional people" (v), thereby continuing the tradition of joining engineering and science with the management systems
that grew from them. So when McDonald said that an "unduly practical" person will suffer "the missing of opportunity and the lack of
support from associates," he suggested that an engineer who was too
rigidly scientific would not do well in business. He elaborated that
point when he stated, "A recent survey of a number of American industrial organizations disclosed the surprising fact that many companies engaged in engineering prefer a president who has been a lawyer
to one who has been an engineer" (157). According to a financial
writer that McDonald cited, lawyers were preferred over engineers
because of two characteristics that lawyers possessed: "a sound
knowledge of the influence of past events on human nature; and the
realization that in a controversy, probably neither side is entirely
right" (157). McDonald asserted that engineers and other people who
were too rigidly scientific did not embody these two characteristics
that would prepare them to be successful managers and potential corporate presidents. McDonald's advice to engineers, scientists, and
business professionals was to read history:

> It is not military and political history that a scientist needs to read in
> order to broaden his mind. Perhaps the most appropriate name for
> the variety best suited to his problem is the history of civilization,
> with a strong leaning to what is more specifically called the history of
> science. . . . They would be conscious of a common background of cul-
> ture and humanism, which would not only weld together the various
> groups of technical specialists but would bridge the gap between
> them and the world at large (160–61).

McDonald advocated a reading list that included Aristotle's works, as
well as biographies of famous scientists and histories of physics, chemistry, transportation, communication, electricity, science, and religion.
He justified this course of reading, as opposed to a more classical or literary reading list, by arguing that reading in the history of science
could provide a shared body of cultural knowledge among scientists in
much the same way that classical education provided a common cultural knowledge among earlier college graduates:

> If culture means broadening the mind and the sympathies, disciplin-
> ing and refining individual tastes, and inspiring enthusiasm for a
> high level of civilization, then the history of science will fulfill the def-
> inition. Such a course of reading supplies the humanistic appeal of
> the old classical training, in addition to unifying for the student much
> of the specialized knowledge gained from more differentiated sci-
> ences (162).

Through this infusion of humanistic histories, the scientific person
could gain the insight to put science into perspective with civilization,
the better to improve that civilization. This course of reading would
also unify scientists through a shared culture and better prepare them
to succeed in business.

McDonald suggested that engineers read histories of science and
civilization because he felt these histories constituted the literary canon
for science. It was for literature students to read "the writings of poets,
novelists, essayists, and dramatists" (162). Engineer John Butler Johnson disagreed with McDonald's literary canon for science in "Two
Kinds of Education for Engineers," included in *Engineering Education:
Essays for English* (1928). Johnson argued instead that engineers should
receive a primarily technical education, but supplement it with extra-
curricular literary readings and cultural activities. Johnson saw "two
general classes of competency which are generated in schools. These are
Competency to Serve, and Competency to Appreciate and Enjoy" (Johnson's
italics, 68). He held that an education for service was the better education for people who aspired to enter business and be productive members of an industrialized society:

> If personal pleasure and happiness are the chief end and aim in life,
> then for those persons who have no disposition to serve, the cultural
> education is the more worthy of admiration and selection . . . If, how-
> ever, service to others is the most worthy purpose in life, and if, in ad-
> dition, such service brings the greatest happiness, then the education
> which develops the ability to serve, in some capacity, should be re-
> garded as the higher and more worthy. This kind of education has the
> further advantage that the money consideration it brings makes its
> possessor a self-supporting member of society instead of a drone or
> parasite, which those must be who cannot serve (69).

In Johnson's dichotomy, students of literature or culture in the classical
liberal arts tradition were parasites on productive society, while students of engineering and science in the modern technical model were

worthy and productive servants of society. Johnson did recognize, however, some limitations for engineers who only received a technical education, i.e., they were not prepared to become well-rounded husbands, fathers, citizens, or neighbors (78), nor were they prepared to become successful managers:

> In order, therefore, that the technical man, who in material things knows what to do, and how to do it, may be able to get the thing done, and to direct the doing of it, he must be an engineer of men and of capital as well as of the materials and forces of Nature. In other words, he must cultivate human interests, human learning, human associations, and avail himself of every opportunity to further these personal and business relations (79).

In Johnson's view, engineers needed to supplement their technical education to make them better prepared to be engineers "of men and of capital" and this was how the humanities could be of use to engineering students. Unlike McDonald, who advised readings only in the history of science, Johnson advised readings in a wide range of disciplines: "For your own personal happiness, and that of your immediate associates, secure in some way . . . an acquaintance with some of the world's best literature, with the leading facts of history, and with the biographies of the greatest men in pure and applied science, as well as with those of statesmen and leaders in many fields" (80). This humanistic supplement to the technical education would enable engineers to become successful managers and enjoy the fruits of their productive lives serving society.

Johnson may have seen a place for the humanities in a technical education, but other engineers did not completely share his view. In the same collection that contained Johnson's essay, James Shotwell presented his thoughts on "Mechanism and Culture."[5] In this essay, Shotwell found a long-standing aversion to mechanical progress among artists and philosophers, going back to Socrates' objections to writing in *The Phaedrus*:

> It is an attitude to be found throughout the whole history of culture. Its most earnest advocates have been the artists, impatient of anything interposed between Nature and the individual. But idealists generally have joined in the denunciation or shared the contempt for mechanism, no matter what their field. Literature has held aloof, except in patronizing, romantic moods, until the present. History has ignored the very implements of progress (202).

Shotwell saw this "contempt for mechanism" as a conspiracy threatening the stability of society, represented by the horrors of World War I:

> There have been, in recent years, some signs of a revolt against the conspiracy of the poetically-minded to ignore the creations of the practically-minded, but unless the revolt becomes a revolution we shall never square ourselves with reality. . . . Idealism, left to itself, is futility. There is no sadder fact in the tragic circumstances of the present than that idealism failed to avert the desolation of Europe. It will always fail, so long as it holds itself aloof from the grimy facts of daily life (203).

If engineers were to serve society and avert future threats from idealistic approaches to world stability, they must put ideas of literature and art to work in mechanisms that ensured progress and stability through science. Art, like nature, was only beneficial for humankind when dominated and controlled by scientists: "Like the forces of Nature, ideas must be harnessed and set to work, or things will remain exactly as they were before" (203). Dante's poetry, for example, could be consecrated as an unchanging object of "fetish worship" (212), or it could be seen in its "immortality of use" within its cultural contexts (213). Shotwell argued for this latter view of studying literature and history within its effects on and uses for a situated culture.

In a companion essay in *Engineering Education*, Bertrand Russell found an inevitable outcome to the influences of industrialization that underpinned Shotwell's argument. In Russell's essay "Where Is Industrialism Going," he held that the effects of industrialism included placing value on the (mass) production of objects—even artistic objects—rather than on the use of those objects:

> The essence of industrialism . . . is an extension of the practice of making tools. In an industrial community the great majority of the population are not making consumable commodities, but only machines and appliances by means of which others can make consumable commodities. This leads men to become utilitarian rather than artistic, since their product has not in itself any direct human value. . . . The man who reads a book is thought to be wasting his time, whereas the man who makes the paper, the man who sets the type, the man who does the binding, and the librarian who catalogues it are all regarded as performing valuable functions. The journey from means to end is so long . . . that people lose sight of the end altogether and come to think more production the only thing that is of importance. Quantity is valued more than quality, and mechanism more than its uses (231).

Russell captured the tension between technical and liberal arts educa-
tions as being part of a larger tension between practices that had been
separated into science and art. Where people considered science to be
utilitarian and art to be the antithesis of science, these people would
value practicality and not recognize the value of impractical art—not
even the artistic uses embodied in the mechanical products of their
utilitarian science. In this industrialized society, people would highly
value the reproduction of large quantities of products, but not recognize or highly value the quality of the many products they reproduced.
In this context, literature and art were only valued, as Shotwell had
argued, for what they could accomplish within a society.

This modern concern for production and practicality was illustrated by T. A. Rickard's reaction to John Masefield's poem "Cargoes"
(1925). Like Shotwell, Rickard argued that romantic idealists mis-
understood the glories of modern science and thus threatened democratic social stability. In *Man and Metals* (1932), Rickard waxed poetic
himself in disputing Masefield's romantic view of the riches of mining:

> A modern poet has contrasted the quinquireme of Nineveh with its
> load of ivory and apes, and the Spanish galleon bringing emeralds
> and gold moidores, with the "dirty British coaster . . . butting through
> the Channel," bearing a cargo of Tyne coal, steel rails, and pig lead;
> . . . as if the useful things on the unromantic vessels were despicable
> as compared with the picturesque ships . . . in days long past. The
> sentiment, of course, is all awry; the quinquireme, rowed by galley-
> slaves, afraid to go out of sight of land, and the galleon, with its
> clumsy sails dependent upon a favorable wind, do not compare in ef-
> fectiveness with the tramp steamer competent to make forceful
> progress in any weather by aid of a tireless steam-engine; and as for
> their cargoes . . . the symbols of savage luxury or barbaric splendor
> . . . they were essentially useless things, whereas the coal is the sign of
> the mechanical energy that has freed the slave and lightened the bur-
> den of the toiler; the steel rails are the symbol of communication and
> commerce of a range and a speed such as the world of quinquiremes
> and galleons never dreamed; and even the lead has its touch of true
> romance when we think of it as a covering for the whispering wires
> that carry the human voice under land and sea . . . It may be that coal
> and iron lack the beauty of emeralds, amethysts, and gold; . . . but as-
> suredly they are emblematic of our civilization, in which the comfort
> of the many is preferred to the glory of the few, in which the useful-
> ness that ministers to human welfare is deemed of more consequence
> than the pomp and glitter of king and conqueror (2:828).

This lengthy quote is included because in it Rickard gathered up many threads of argument in the contest between making knowledge through the liberal arts or through the sciences. In the beginning of the passage, engineer and scientist Rickard set himself in opposition to the "modern poet" whose poem will provide an example of unscientific thinking. The poet believed the ancient ships filled with valuable gems and gold are romantic ideals to be sought in a modern world. The poet set these ancient ideals in contrast to "dirty" modern ships full of base coal, steel, and lead. Rickard pointed out that the poet found the modern ships "unromantic" and their cargo "despicable," but Rickard found this poetic point of view to be "all awry." Instead of valuing the romantic ancient ships and their precious cargoes, Rickard found that they represented ignorance, incompetence, and despotism. For the modern engineer, value was in the potential for "mechanical energy" embodied in the coal, steel, and lead, which could build a democratic society concerned with the welfare of the masses. The base metals in the modern tramp steamers certainly had more value, both monetarily and culturally, than the gems and precious metals of the ancient ships.

For Rickard, the ancient ships and their cargoes were powered by slaves and uncontrollable natural winds. The captains of these ships, without sophisticated navigational instruments, dared not sail out of the sight of land lest they lose their way in uncharted seas. The cargoes were "essentially useless things" fit only for savages and barbarians, not for modern civilization. Modern civilization, on the other hand, was symbolized by the coal, steel, and lead on the British coaster. Unlike gems and gold, these modern riches from the mines "freed the slave and lightened the burden of the toiler" through their applications brought about by engineers using scientific knowledge. Because of scientists and engineers, steel could be rolled into durable rails on which train cars, powered by coal-burning steam engines, sped goods and passengers in commerce over long distances. Lead could be used to cover "whispering wires that carry the human voice" doing business well beyond earshot. Because of scientists and engineers, savage and barbaric civilizations of ancient times were transformed into a modern civilization "in which the comfort of the many is preferred to the glory of the few" and despotism can be replaced by democratic rule.

In Rickard's romance of science and engineering, modern civilization depended on the "dirty" ships and their "despicable" cargoes. In upholding ancient ideals, Masefield showed himself to be in favor of savagery, barbarism, and despotism. In the impractical poet's world, according to Rickard, the welfare of the many was sacrificed for the benefit of the few and useless gems and gold were valued more highly than the metals that enabled modern machinery to improve the general living standards. Clearly, Rickard felt that literature, which romanticized what he considered to be a barbaric past, was not fit reading for the engineers who would shape a better future through applied science. He would probably not agree with English professors Crouch and Zetler's choice of including John Masefield's poems on a reading list for engineering students.

TECHNICAL WRITING MOVES
FROM ENGINEERING TO ENGLISH

In the midst of this tension between engineering and literary studies, technical writing instruction within the academy was moving from engineering and the sciences into disciplines traditionally allied with the classical liberal arts. Concurrently, technical writing practice was moving out of engineering and into a separate specialization, much as communication in industry had moved from being a management function to a clerical specialization early in the 20th century. The separation of technical writing from engineering took place during and shortly after World War II, as Jay Gould described in his 1964 edition of *Opportunities in Technical Writing*:

> World War II is an important date for the technical writing profession for at that time the great industries involved in turning out guns, planes, and chemicals came to realize that communicating was nearly as important as manufacturing all the necessary things for the war machine. Reports had to be written for the men and women who were inventing the machines and the electronic systems . . . The engineers and scientists were just too busy to do all this writing themselves. They had more important jobs to do (12).

Earlier in the century, the role of clerical labor within systematized management had been transformed into lower-paid help to relieve higher-paid managers of the burden for generating communications that controlled the system. So the role of technical writer was transformed into lower-paid help to relieve higher-paid engineers and scientists of the burden for generating communications that stabilized the dominance of scientific knowledge in our industrialized society. Like the increasing paperwork that made paying clerical workers more profitable than using managers' high-priced time for generating communications, the burden of increased technology development made paying technical writers profitable. According to *Technical Writing* authors Wellborn, Green, and Nall, the demand for technical writers by 1960 had grown in importance and in numbers: "This ever-increasing demand is the necessary result of accelerated experimentation in science, agriculture, engineering, and industry. Technical communication is the indispensable clearing house for information and ideas in an era of rapidly accumulating knowledge" (4). Engineers and scientists could no longer profitably spend their time writing reports when they could be developing technologies to improve general living conditions. The role of the technical writer, therefore, was separated from that of the scientist and an opportunity was created for specialized writers to take over the communication function for engineers and scientists.

For some time before World War II, technical writing was seen as something other than the average English composition course. Philip McDonald, for example, disparaged English composition in the preface to his technical writing textbook *English and Science* (1929):

> In this modern age of science and industry we are compelled to think clearly and to speak and write concisely. A student of science, in particular, should express himself in as clear and concise a style as possible. Yet much of the old-fashioned rhetoric and composition taught in our schools and colleges encourages verbosity and tolerates vague, stilted language. Such teaching is not suitable for the education of engineers, scientists, and most business and professional people. . . . The most important officials in modern civilization usually write only two kinds of composition—letters and reports (v).

As much as McDonald railed against the unsuitability of English composition courses for technical education, his textbook approached writing instruction in much the same way as English composition textbooks had for nearly 50 years. In focusing on the forms of reports and letters, for example, McDonald followed in the composition tradition of categorizing forms of writing. John Quackenbos, for example, used this categorizing approach in *Practical Rhetoric* (1896), when he covered the "Standard Prose Forms" (8) of letter, essay, narrative, biography, fiction, novel, and sermon. Similarly, Charles Baldwin set out the following literary forms in his textbook *Composition Oral and Written* (1909): epic, romance, lyric, drama, oratory, essay, novel, short story (xii–xiii). These forms are related to the modes of discourse that Robert Connors, in "The Rise and Fall of the Modes of Discourse," traced through Samuel Newman's *A Practical System of Rhetoric* (1827) and

Alexander Bain's *English Composition and Rhetoric* (1866). By the 1890s, these modes had been solidified as narration, description, exposition, argumentation, with persuasion either listed as a fifth form or being subsumed by argumentation. Albert Kitzhaber found these forms of discourse to be the major organizing principle for composition textbooks from the 1890s through the 1930s: "It was in the 1890s that the 'Forms of Discourse' . . . finally triumphed and became the dominant organizing principle behind most of the textbooks in rhetoric" (127). Given the ubiquity of the forms as the organizing principle composition instruction, it is not surprising that technical writing textbook authors, while decrying the sorry state of English composition, should emulate composition's forms of discourse in presenting technical writing instruction.

Of the four or five forms that were typically used to categorize written discourse, exposition was the one form that addressed scientific or practical information. In *Practical Rhetoric* (1896), Quackenbos quoted Professor Bain's definition of exposition in an epigraph beginning his chapter on this topic: "Exposition is applicable to knowledge or information in the form of what is called the sciences, as mathematics, natural philosophy, chemistry, physiology, natural history, the human mind" (130). Exposition, therefore, was the form most suited to the practical purposes of technical writing. Nan Johnson argued that because of this emphasis on practicality,

> exposition subsume[d] all forms of objective analysis: textbooks; . . . practical instructions regarding machinery and industrial processes; instructions regarding household skills and agricultural methods; scientific treatises on natural and mental phenomena . . . and critical explanations of literary and artistic principles. . . . Considered the predominant form of prose in general use, exposition enjoyed the reputation among nineteenth-century and early twentieth-century rhetoricians as the 'workaday form' of composition (210).

According to this definition, the practical letters and reports covered in technical writing texts were subcategories of the dominant mode of discourse—exposition. Technical writing also covered a second form of discourse—description—because, as Quackenbos found, "Clearness of vision is at the foundation of a talent for description" (93). John Genung defined description in *The Working Principles of Rhetoric* (1901), saying that it "deal[s] with observed objects" (475). In a culture of empirical science, the ability to clearly observe and clearly communicate observations was an important technical ability. In technical writing textbooks, however, description often was subsumed under the dominant category of exposition.[6]

If exposition was the dominant mode of discourse in technical writing, clarity was the dominant quality of style throughout the 20th century. In Rickard's 1910 technical writing textbook, he counseled that "precedence must be given to clearness of statement. . . . Be lucid, and all these other qualities shall be yours, as you desire them and practice to attain them" (84). McDonald gave similar advice to technical writing students in *English and Science* (1929): "The simple, concise style is best for most individuals" (89). Harwell advised in *Technical Communication* (1969), "To be clear should be our constant aim" (3). And further, "Clearness and simplicity are practically inseparable virtues. That which is clear is usually simple in structure and diction, and that which is simple can nearly always be understood quickly" (9). John Lannon defined clarity in *Technical Writing* (1994) in much the same way: "A clear sentence conveys the writer's exact meaning on the first reading" (251). This focus on clarity in technical communication works to render language as the invisible conduit through which scientific knowledge is carried from one mind to another. This metaphor of language as an invisible conduit enables science to have a material reality untouched by language as its object. As important as this metaphor is for technical writing, the same emphasis on clarity can be seen in composition studies at the end of the 19th century.

John Genung held clarity to be the most important aspect of style in his composition textbook *Practical Elements* (1886): "The first and indispensable quality of a good style is Clearness. . . . Nor is it enough for perfect clearness that a style be intelligible. Quintilian puts the ideal still higher. . . . not language that may be understood, but language that cannot fail to be understood, is the writer's true aim" (19). Composition instructor Barrett Wendell described clearness in much the same terms in *English Composition* (1891): "To be thoroughly clear, it is not enough that style express the writer's meaning; style must so express this meaning that no rational reader can have any doubt as to what the meaning is" (194). Baldwin, too, found clarity to be at the heart of composition in *Composition Oral and Written* (1909): "To be clear, to be interesting, are objects always; and all the ways of gaining them spring from that root idea of adapting oneself to readers or hearers" (2). So pervasive was this emphasis on clarity within composition that Nan Johnson argued,

> Perspicuity, often defined as "clearness" or "clarity," was regarded by
> nineteenth-century rhetoricians as a necessary condition for "the full

effective transfer of thought." In order to achieve this "transfer," the writer never presents the reader with a word or construction that requires undue scrutiny or that creates confusion regarding what meaning is intended (192).

Composition at the turn of the 20th century was as concerned with clarity of expression as technical writing was then—and continues to be at the turn of the 21st century.

A third concern shared between technical writing and English composition was the use of correct English. Rickard's emphasis in his *Guide* (1910) was on the use of correct grammar, word choice, and standardized mining terms. McDonald began his chapter "Correct Language, the Basis of All Writing" (1929) with the advice, "In order to write satisfactory reports and letters, it is necessary to understand language and to know how to use it properly. One of the most important tests of an educated man should be the question, can he express his ideas satisfactorily?" (83). Crouch and Zetler, too, connected grammatical correctness to the important quality of clearness in the beginning of their chapter "Language Essentials" (1948): "Grammatical accuracy is necessary if the communication is to be clear" (273). These authors also included a chapter on usage rules as a further resource for correct grammar. Harwell included a similar section on grammar and usage at the end of the 1969 edition of his textbook.

This concern with correctness in technical communication mirrored a similar and long-standing concern in English composition. For example, in *English Composition* (1891) Barrett Wendell cautioned students, "All offenses against good use in our choice of words are either Barbarisms or Improprieties" (44). Wendell argued that through the use of correct words, students could strive for an ideal "in the use of words as in other phases of our conduct of life" (75). In *Practical Rhetoric* (1896), Quackenbos agreed that students should strive for an ideal of language and life through the use of correct words. He advised students to choose words that were in standard usage according to three criteria: "Reputable, that is authorized by the majority of writers of high reputation . . . National, as opposed to provincial and foreign. . . . Present, as opposed to past or probable future. The reputable national use of one period differs materially from that of another" (137–38). Quackenbos further held, "Good Use employs Pure Words.—Purity, as a quality of style, implies the use of English words in authorized senses" (142). In his extensive chapters on word choice, Quackenbos advised against using words that were obsolete, "Words foisted into the Language by Irresponsible Inventors" (146), "Hybrid or Mongrel Words" (147), compound words,

"Foreign and Local Words [that] are not National, and hence not English" (149), colloquialisms, vulgarisms, slang, and redundancies. This excessive concern with correct language reflected social conditions in which middle-class people, working in U.S. industries stabilized through systematized management, strove to better their condition through college education, important social contacts, and correct behavior. Michael Halloran found that this concern with upward social mobility gained importance in the late 19th century: "In the competitive middle-class society of the nineteenth-century, speaking and writing 'correct' English took on new importance as a sign of membership in the upper strata. The new reliance on writing as a medium of evaluation lent further importance to correctness" (167). Robert Connors also found this striving for upward social mobility to be an important factor underlying the concern with correct English. In his article "The Rhetoric of Mechanical Correctness," Connors cited an additional factor in this concern with class distinctions:

> This new interest seems to have sprung from two distinct proximate causes: the Eastern reaction against the "roughness" and "crudeness" of frontier America, an attitude which wished to set standards of propriety in language as in all other aspects of life; and the desire for self-improvement and "getting ahead," which was an important part of the American mythos during the nineteenth century (30–31).

This concern with the connection between upward social mobility and correct English is illustrated by practicing engineer John Harrington who, in "The Value of English to the Technical Man" (1928), explained why engineers should be concerned with correct English: "One who is thus ignorant of the language finds social progress substantially impossible. . . . Matters of very large importance are often settled by favor, and favor frequently follows social position" (96). Students of technical writing, like their counterparts in English composition classes, were interested in the personal benefits they could earn through their education in correct language use. If students learned to accumulate the cultural capital that could be gained through using correct language, they could reap rewards such as elevated social standing, professional promotion, and increased salaries. Engineers at the beginning of the 20th century were seeking to elevate their professional standing through connecting their work with pure science. They were also seeking to elevate their social standing through the use of correct English. Thus, instruction in this topic appeared in technical writing textbooks as well as in composition textbooks.

With so many interests in common—exposition, description, clear style, correct usage—it is not surprising that technical writing instruction migrated from the scientific realm of the engineers to the literary realm of the English instructors as technical writing practice became more specialized and separated from engineering practice. Perhaps this migration fueled tensions between scientific and artistic knowledge that were explored in the previous section of this chapter. After all, as Rickard set the argument early in the century, "[W]riting is an instrument for transmitting ideas" (*Guide* 9) and if the scientist's ideas are not transmitted, they are not knowledge. Without technical writing—the currency of science—the scientist's ideas are worthless. A scientist's ideas may have an illusory stamp of science within her mind, but there is no metal coin to bear that stamp until the scientist's knowledge takes material form through communication. Without technical writing, science is a pauper with no prospects for upward mobility. The question of who controls technical writing is really a question of who controls the cultural capital and material value of scientific knowledge.

Perhaps when technical writing separated from engineering in a division analogous to clerical workers separating from managers, engineers and scientists were lulled into complacency by thinking that they, too, would have a malleable group of subordinates—technical writers as stenographers—to take their dictation and transcribe their words. Perhaps they were flattered by their relative elevation in position resulting from the installation of a group of workers below them. After all, this addition of clerical support staff had worked well for the engineers who became managers some years earlier. Why not let the English Department see to the training of this engineering support staff? In the academic system, the science and engineering departments can better use their time developing scientific knowledge and technological applications than teaching engineering support staff to write. By training technical writers, English Departments could put their literary idealism to practical use and become productive members of society, contributing to a stable progress and improvement of humankind instead of simply being parasites reading books.

By the mid-1950s, however, the new occupation of technical writing was rapidly organizing and expanding. Jay Gould, who played an important role in the development of this profession, described these early years in the field:

By March, 1955, technical writers from the Boston area had organized themselves to such an extent that they felt the time had come to draw up a constitution. This was the Society of Technical Writers, which by that time had informal chapters in Connecticut, Illinois, Maryland, Massachusetts, New Jersey, New York, Pennsylvania, and Tennessee (59).

Other technical writing associations came together in New York City, and on the West Coast. By the mid-1960s, Gould stated that "the official professional society of the technical writing profession is the Society of Technical Writers and Publishers" (62), with "founder chapters in Boston, New York, and Los Angeles. Chicago, Cleveland, and Washington have vigorous chapters, as do Pittsburgh and Cincinnati. Some chapters combine cities and districts, as the Dayton-Miami Valley area of Ohio and the Rocky Mountain area" (63). Clearly, the technical writing profession grew rapidly after World War II. Gould listed 36 two- and four-year colleges offering at least one course in some type of technical writing. He also listed four universities with full curricula in technical writing: Colorado State University, Division of Technical Journalism, Department of English; Margaret Morrison Carnegie College, "the women's division of Carnegie Institute of Technology, offer[ing] a four-year program in technical writing and editing" (99); Illinois Institute of Technology at Chicago offer[ing] "two programs: one in Science Writing and one in Science Information" (101); and Rensselaer Polytechnic Institute, where Gould had developed a "two-semester curriculum leading to the degree of master of science" (104), concentrating in communication theory or technical writing. This expansion of technical writing instruction fueled the need for technical writing textbooks, which now resembled English composition textbooks more than engineering handbooks.

In 1954, Gordon Mills and John Walter authored *Technical Writing* following the new model of technical writing practice. Instead of instructing engineers in managerial communication skills, as Crouch and Zetler had done in 1948, Mills and Walter sought to redefine technical writing as a specialized practice of its own. In their preface to the first edition, they described their initial uncertainty about the boundaries of technical writing instruction: "The difficulty was partly that the limits of the subject were uncertain; apparently nobody had ever seriously explored the concept of technical writing with the purpose of trying to say precisely what technical writing is" (viii). Since technical writing was a new profession, it is understandable that nobody had yet mapped its boundaries, especially in relation to the engineering and scientific professions. Technical writing was formerly contained within the boundaries of engineering and science. Now Mills and Walter

sought to draw new boundaries around technical writing as a specialized practice separate from—but related to—engineering and science.
They charted their course in a scientific way:

> [W]e undertook three investigations. We began by seeking examples
> of reports and articles, and expressions of opinion about important
> problems; altogether we incurred an indebtedness to over three hun-
> dred industrial and research organizations in making our survey. We
> also worked out, in writing, a theory of what technical writing is . . .
> Thirdly, we studied the content and organization of college courses in
> the subject (vii–ix).

In addition to their scientific investigations, Mills and Walter claimed
authority in the field by referring to their experience teaching at the
University of Texas and working in industry. They also acknowledged
their debt to the people they consulted at General Electric, the Civil
Aeronautics Administration, General Motors, and The Texas Company (x). This recitation of their investigations, experience, and industry contacts served to legitimate Mills and Walter as experts in the
new field of technical writing, although they were professors and not
engineers.

   In organizing the contents of their textbook, Mills and Walter drew
on conventional topics from composition and business communication
textbooks as far back as the 1890s. They covered style in Chapter 3, including advice on "Reader Adaptation . . . Precision in the Use of
Words—Sentence Structure . . . Paragraph Structure . . . Grammar and
Usage—Common Errors in Usage—Mechanics of Style—Effective
Word Choice Exercise" (xi). In Chapters 5 through 9 they covered a
version of the traditional forms of discourse: definition, description,
classification, and interpretation (xii). In Chapters 13 through 19 they
covered the technical subcategories of exposition: reports, letters, and
articles (xiii). In Chapters 20 and 21 they covered formatting and
graphical illustrations in much the same way Hotchkiss and Kilduff
had covered business letter formatting in their *Handbook of Business English* (1917) and Carey had covered this same subject in *Business Writing* (1949). Mills and Walter then set out the library research process in
Chapters 22 and 23, much as Baldwin had covered collecting and
grouping facts in *Composition Oral and Written* (1909). Taken as a whole,
Mills and Walter's approach to technical writing instruction was heav-
ily influenced by traditional English composition approaches to writing instruction, placing their textbook solidly in the classical liberal

arts tradition, despite their scientific investigations to delimit the new field of technical writing.

Mills and Walter's reasserted their sympathy with the humanities in their 1970 preface to the third edition. Here the authors stated that they had revised their book in light of

> particularly interesting changes recent years have seen in attitudes to-ward language. Developments that have been occurring over a long period of time in linguistics, in the philosophy of science, and in the profession of technical writing are now making themselves evident in the classroom and in the general practice of technical writing. . . . sci-entists and philosophers have been cultivating a new and, as it seems to us, a more realistic and genial concept of science. The scientist's role as a person is heightened, and some of the intuitiveness and en-thusiasm of art are accepted as a natural part of his science (v).

In claiming that scientists are accepting the "intuitiveness and enthu-siasm of art" in their practices, Mills and Walter described a trend in science that, while faithful to the practice of scientists, called into ques-tion the knowledge-making power of the scientific method. In effect, Mills and Walter asserted liberal art's encroachment into scientific ter-ritory, a development of technical writing becoming a specialty within the humanities. When Mills and Walter said that developments in lin-guistics and the philosophy of science were changing scientific knowl-edge, they reminded scientists of the control technical writing exercises over scientific knowledge. They illuminated the knowledge-making power of communication. They highlighted the control communica-tion exercises over systems of knowledge and power and the loss of control that engineering and science could suffer by relinquishing the practice of technical writing. They foregrounded an incipient threat to the stability of cultural relations between science and art—relations tortuously overturned and stabilized throughout hundreds of years in contests between science and speculation. The stamp of science, which gave technical writing its value, would not be replaced by the stamp of humanistic speculation.

ENGLISH EMBRACES SCIENCE

Since the 1890s, English composition instruction had been influenced by trends toward viewing the body of composition knowledge as a scientific system controlled by standards and laws. The concern with correct English, discussed in the previous section of this chapter, can be seen as one such move toward standardizing language into a system closely related to systematic management and the personal and social benefits promised to those workers who successfully internalized that management system. Professional workers who spoke and wrote correct English, for example, would be eligible to move up through the management ranks, where pay and social standing were higher than manual workers could expect. In his 1896 composition textbook *Practical Rhetoric*, Quackenbos sought to improve the study of rhetoric in response to Herbert Spencer's earlier attempt to systematize this discipline in his "Essay on the Philosophy of Style." Quackenbos first noted that Spencer "observes that the maxims contained in works on composition and rhetoric are presented in an unorganized form, and proceeds to systematize the scattered precepts under one leading principle, economy" (3). Quackenbos then attempted to advance Spencer's system by finding "an answer to the question, Is further generalization possible?" (3). In other words, Quackenbos wanted to conform composition instruction more closely to scientific models of law-governed practice than Spencer was able to do with his overarching law of economy. Instead of economy being the principle law of composition, Quackenbos argued that "all rhetorical law [derived] from that principle of beauty known as harmony, or adaptation,—a principle which includes economy, as well as order, unity in variety, and proportion" (3). In claiming that harmony was a more appropriate law than economy to govern rhetoric, Quackenbos argued that a humanistic study such as rhetoric could also be considered a science.

Quackenbos clearly stated his view that rhetoric or composition was a science in his heading to Lesson I: "Sciences Subsidiary to Rhetoric" (19). In this section, the author claimed that rhetoric "assumes an acquaintance on our part with certain great fundamental sciences. These are: Grammar, which enforces correctness . . . logic, which determines the laws of intellect . . . ethics, the science of morals . . . and æsthetics, the science of beauty" (19). In Quackenbos' system, rhetoric or composition was based on these four fundamental sciences, which were themselves based on other fundamental laws. Aesthetics, for example, was based on "fundamental truths of beauty." These "truths of beauty" underlying all rhetoric were scientifically derived from a naive Baconian investigation of great literary masterpieces: "We have seen that the laws of rhetoric were induced from a critical study of many masterpieces,—literary works which elicited universal admiration, and which had been constructed intuitively in accordance with great psychological principles"(20). In keeping with his psychological frame-work, Quackenbos then explained that the appreciation of beauty re-sides in "taste," which for him was a physical phenomenon:

> The word *taste* is derived from a root (tag) meaning "to touch with the fingers." It was used secondarily to explain the touch of the special nerves of the tongue, and finally to designate the action of the mind itself in touching or feeling the beauty of things. It thus describes the æsthetic faculty, the power of discerning beauty, and of deriving pleasure both from the act itself and from the qualities perceived (Quackenbos' italics, 20–21).

In constructing taste as a physical phenomenon and aesthetics as gov-erned by psychological laws, Quackenbos placed aesthetics, rhetoric, and composition in the realm of the sciences. Because psychology was closely tied with physiology at the turn of the century, Quackenbos' ar-gument worked to make composition a branch of the science of physiology. Ross Winterowd explained the importance of Quackenbos' argument for shaping composition as a scientific language practice:

> The nature of a field given to the study of language (i.e., linguistics) will depend on the definition of language accepted by practitioners in the discipline. If "language is in the main muscular movement, either voluntary or involuntary, determined by changes in the nervous system of the communicator" (Walter G. Pillsbury and Clarence L. Meader, *The Psychology of Language* [New York: D. Appleton and Company, 1928], pp. 4–5), then linguistics will ultimately be a branch of physiology (265).

Not only did Quackenbos place rhetoric and composition in the realm of science by claiming that aesthetics and rhetoric relied on a physical interpretation of taste, he also described a useful role for rhetoric and composition that would have probably pleased even John Shotwell. Composition would train students to be critics: "The student must learn to form literary judgments, to value correctly, as well as to pro-duce, creditable literary work. If he be inspired with a true sense of the importance of rhetorical study, he will desire from the outset, not only to write, but also to judge" (13). In a culture where scientific progress resulted in rapid social change, composition instruction could teach students to use correct English and evaluate the English used by other people. In this way, upward social mobility would be assured for those who used correct language, as scientifically observed by those who used correct language.

This idea of the composition teacher as language expert or linguist, scientifically observing phenomena governed by laws of psychology or physiology, persisted well past the middle of the 20th century. Nelson Francis, for example, included the following description of a linguist in *The English Language, An Introduction: Background for Writing* (1965):

> From the point of view of the objective student of language, the linguist, the grammar of a language is a description of the ways in which the language uses patterns of structure to convey meaning. In this sense, then, the student of grammar is an observer, an analyst, and a maker of descriptive formulas. The fact that these formulas are sometimes called rules or laws often leads people to think of the grammarian as a legislator. But he is not. His rules are like the "laws of nature" that scientists talk about; they are generalized formulas based on the way things are and happen (8).

The rules of language might be "like the 'laws of nature,'" but they are not really laws in a strictly scientific sense. Kenneth Houp and Thomas Pearsall distinguished between laws of nature and laws of language in their 1968 textbook *Reporting Technical Information*:

> As a technician, engineer, or scientist—who must sometimes write— understand that writing is something that must be learned, even as chemistry, physics, and mathematics must be. The rules and formulas of writing are not as exact, perhaps, as those of science, but they can never be thrown overboard if you are to bring your substance to your reader (5).

Technical communication and English composition can aspire to the status of scientific disciplines, much as middle-class workers aspired to upward social mobility through the use of correct English. But unlike some successful engineers who became managers, technical communication could not really become a science. Technical communication could carry the stamp of science, but it could not be a science itself. It could be the coinage of science, but only as metal for the coin. The value of the coin resided not in its metal, however, but in its potential for exchange as knowledge with the stamp of science circulating in an economy based on science and technology. No matter how much technical communication and English composition tried to distract attention away from their humanistic origins and to attract attention to their science-like laws, their origins could not be buried. Mills and Walter pointed out the relationship of technical writing and science in describing technical writing as being characterized by a "scientific point of view" (4). Technical writers were not scientists, but they could think like scientists.

By the mid-1950s, Jay Gould had worked with other technical writers to professionalize their practice and Mills and Walter had followed that trend by authoring a textbook that was suitable for training technical writers in traditionally liberal arts-based university departments. Two decades after these developments, however, J. C. Mathes and Dwight Stevenson published a textbook in the traditional engineering-based model of technical writing. In their preface to *Designing Technical Reports* (1976), Mathes and Stevenson first put forward the traditional apology for studying technical writing, i.e., personal financial benefit: "An engineer in management asserts, 'an engineer who can't communicate is in trouble; a manager who can't communicate is finished.' Another engineer warns, 'if you can't tell them what you are doing, they'll get someone else to do it'" (xv). This familiar argument had been used by technical writing textbook authors from T. A. Rickard forward, by business communication textbook authors, and by composition instructors with their concerns for correct English. Mathes and Stevenson also echoed standard technical writing and business communication textbook arguments that English composition instruction was not good preparation for practical writing in business and industry:

> Yet engineers are not trained well by their college writing experiences. Despite the usual freshman English themes on "Justice" or "Civil Disobedience," the term papers in history or American literature, the numerous lab reports, and despite the occasional technical report or design project report, engineering graduates are usually ill-equipped to cope with the communication needs of professional engineers (xv).

This version of the argument blaming English composition for being inadequate to practical communication needs goes further than earlier arguments. It also blames the academy in general, including science courses, for neglecting practical communication instruction. Mathes and Stevenson explain that in composition courses, students "were asked to write essays. For the most part, though, the purposes of these essays were not to really communicate; the purpose of this type of writing was essentially humanistic. That is, you wrote to improve the ability of your reasoning and the quality of your perceptions and experiences"

(3–4). Here Mathes and Stevenson distinguished between reasoning, perceiving, and experiencing in the humanities, and analyzing and observing in the empirical sciences—practices that could just as easily be seen as more similar than they are different.

Writing using these observational and analytic skills in the humanities did "not really communicate," yet writing in academic science did not fare a much better judgment from these authors:

> As engineers you have had other, somewhat more appropriate, types of writing experiences. . . . However, even these experiences contain pitfalls that continue to trap engineers when they start writing technical reports on the job. Many of your writing activities as an engineering student lacked an appropriate communication purpose. The "system" survives if a student produces a poorly written laboratory report; the only thing that happens is the student flunks (4).

This concern with system was paramount in Mathes and Stevenson's technical writing textbook for engineering students:

> We assume that if an engineer can use rigorous, systematic procedures to resolve technical problems, he or she should be equally able to apply systematic procedures to resolve communication problems. . . . The purpose of this book, therefore, is to present a systematic procedure which will enable the engineer to approach and solve the problem of report design confidently and effectively (xvi).

In viewing technical writing systematically, Mathes and Stevenson defined a technical report in terms reminiscent of records in Taylor's scientific management system: "The technical report is *an act of communication by a professional in an organizational system to transfer information necessary for the system to continue to function*" (Mathes' italics, 3). They further placed the system's needs at the core of technical writing: "The influence of the system is the primary factor to consider when you design a report" (5). For Mathes and Stevenson, the engineer's job was to serve a management system and ensure its survival, thereby ensuring continuing progress, improvement in the general living conditions, and a stable democracy. In return for this service to a management system and humankind, the engineer/writer would gain personal financial rewards.

The relationship between engineering, technical writing, management systems, and society that was developed at the turn of the 20th century was reconstructed in Mathes and Stevenson's 1976 textbook.

Technical writing was not seen as a specialized practice apart from engineering. Legitimate knowledge was not made outside of science, in the classic liberal arts-based curricula of English composition. People trained in English did not communicate knowledge on behalf of mute scientists.

An important question to ask here is, "Why did Mathes and Stevenson's textbook pressure engineering and systematic management to reclaim technical writing?" One answer to this question would be that engineering and systematic management could lose control of scientific knowledge-making by separating technical writing from engineering. As long as engineers wrote about their technological developments, they were the scientific knowledge-making agents in a culture dominated by scientific knowledge. But when engineers relinquished their agency in this transaction, they became powerless, like the unsuccessful engineers who could not communicate well and, therefore, could not progress in the management system to positions of greater status and higher pay. Technical writing, business communication, and English composition textbooks had warned against this unhappy fate since the beginning of the century.

Another answer to this question is caught in the venerable tension between science and speculation as knowledge-making activities. At the end of the last century, mechanical engineers designed the functional-military management system upon which rested 20th-century industrial success in the United States. Up until World War II, engineer/managers had maintained this system by controlling the communications, which, in turn, controlled the system and its workers. After the war, though, managers continued to control communication and information within increasingly computerized management systems, but engineers had taken themselves out of the loop by relinquishing their communication function to specialized technical writers. Unlike the clerical workers who helped managers generate communications, however, technical writers did not only transcribe engineers' words. They gained knowledge of engineering practices and invented their own words to define and describe the technology that engineers developed. By generating communication, technical writers extended their ability to control the management system while engineers lost some of that ability.

Losing control of scientific knowledge to technical writers posed a threat to the dominance of scientists and engineers in our culture. If technical writers could have been trained only in the sciences, this knowledge-making dilemma might have been less acute. But writing instruction had deep roots in liberal arts and the scholasticism that

Frances Bacon and like-minded thinkers overturned with so much difficulty. This displacement of the liberal arts tradition by science was so
thorough that by the late-20th century, we do not really know how to
integrate speculation into knowledge-making activities dominated by
the scientific method, except to keep them at the margins. Because
technical writers have an inescapable liberal arts history, they are not
fully integrated into the scientific practices about which they write.
They are on the margins of this activity in a supplementary position,
yet their function is the very stuff of knowledge-making. Without communication, scientific knowledge does not exist materially in a culture.
Without technical communication, the coin of science does not exist.
And to be authentic coin, the stamp of science must be impressed on
the communication, not the stamp of speculation. The dilemma for
technical writers is how to transform their spurious stamp of speculation into a genuine stamp of science, thereby ensuring stability for the
system based on scientific knowledge.

The transformation that technical writers work is analogous to the
transformations described by Keith Hjortshoj in "The Marginality of
the Left-Hand Castes." One of these transformations described by
Hjortshoj is carried out by the Kammalan caste in southern India,
whose "name ['one who gives the eye'] derives from the fact that Kammalan smiths and stone carvers create the statues of deities installed in
Hindu temples" (496). Giving an eye to a statue also gives great trans-
formational powers to the Kammalan: "This event has unparalleled
importance in the consecration of a sacred image, because a piece of
stone or metal actually becomes a sacred presence only at the moment
at which the Kammalan 'opens its eyes'" (496). Yet the Kammalan hold
a marginal place in Hindu culture. This paradox of power and marginality can be explained by the Kammalan's potentially destabilizing
knowledge concerning the origins of the sacred:

> In symbolically powerful ways, the Kammalans bring the gods and
> goddesses into the life and light of Hindu worship. Or, by inversion,
> artisans represent the potentially destabilizing knowledge that wor-
> shippers of the deities are supposed to forget: knowledge that these
> sacred images were fashioned from stone or metal, by human hands
> (496).

Like the Kammalans who transform stone and metal into sacred objects, technical writers transform mute science into scientific currency
through a craft based in speculation. Like the Kammalans, technical
writers—whether they are scientists, engineers, or specialized writers—

know that scientific knowledge is fashioned by human hands. Yet this knowledge of the origins of science threatens to destabilize the system based on this knowledge. Hjortshoj argued that the people who specialize in this dangerous knowledge are more safely located at the margins of the system:

> Most of us have also noticed . . . an underlying fear—not fear of us, exactly, but of language itself and of its potential to make and destroy careers, redefine the centers and boundaries of disciplines, disrupt or threaten the status quo, and define what is currently true or false, marketable and unmarketable, fashionable or unfashionable, impressive or humiliating. Like fire, language is essential, transformative, and potentially destructive. . . . Keeping us in our place—in the marginal, parenthetical relation to the rest of academic life—is a way of keeping the potentially disruptive power of language contained and disguised, though not altogether denied (500–01).

Like the English composition instructors who prepared the way for technical communication, specialized technical writers and the discipline of technical communication are more safely located at the margins of our culture based on scientific knowledge and systematized management. At the margins, technical writer's experiences with the fragility of scientific knowledge, the human origins of management systems, and the tenuous nature of systematic control based on communication can be overlooked more easily by people whose livelihoods depend on the stability of this knowledge/power system.

8

Whose Knowledge Is Powerful?

Of course, it is not true that we each have equal power to choose our own technology in every case, nor that our reasons for choosing one over another are always good, nor that we always foresee and desire every consequence of those choices. No aspect of life is so simple or straightforward. I do believe, however, that those with sufficient foresight and power over their own affairs create or choose those technologies they think will preserve and increase that power. We might say that these technologies are socially constructed for certain purposes. The impact they have on any one of us is the result of a complex calculus of class, race, gender, luck, and other similar variables. There is certainly a sense in which we all—perpetrators, victims, beneficiaries, bystanders—collaborate in this social construction. But it is equally certain that we do not all have an equal say in that activity.
— CARROLL PURSELL, *The Machine in America,* 1995

This study ends as it began, asking Foucault's question, "How is it that one particular statement appeared rather than another?" (*Archaeology* 27). In the discourse of 20th-century technical writing in the United States, we begin to see that particular statements appeared as knowledge because they could bear the stamp of science. Other statements did not appear as knowledge because they bore other stamps, such as the stamps of speculation, of liberal arts, of classical knowledge, of scholasticism, of opinion, of religion, of alchemy, or of magic. Of course, this situation is particular to the 20th century in the United States. In 16th-century Germany, for example, statements appeared as knowledge that bore the stamps of religion, of scholasticism, or of classical knowledge. Other statements based on experience or experimental observation were not considered to be legitimate knowledge. What we now know as science, in other words, was not legitimate knowledge four centuries ago.

A question for cultural researchers of technical communication, therefore, is "Why is scientific knowledge legitimate at this time and place while other kinds of knowledge are marginalized or silent?" This study begins to explore this question by relating the development of technical writing practices to developments in engineering and management over the course of this century. Even this confined focus deserves a more complete study than was possible in this preliminary history. Many intellectual trends merely mentioned in this current study can yield further insights into the role of technical writing in our culture. The relationship between management and military science, for example, needs further elaboration to explore how technical writing upholds militaristic concepts of industry in our country. Early management engineer's reliance on 19th-century Prussian line-and-staff military organization provides an especially rich lode of knowledge development.

The development of textbooks as cultural artifacts is another topic that was touched on in this study, but that bears further research. The idea of textbooks as compilations of knowledge was explored in a cursory fashion here. But its complementary idea of textbooks as copybooks was omitted entirely from this discussion for lack of space and time. Combining these two complementary ideas of the textbook is crucial to a more complete understanding of how compilation textbooks work to stabilize legitimate knowledge. Only by exploring how copybook textbooks can destabilize knowledge—through copying errors, through the writer's choice of passages to copy, through the teacher's choice of passages and marginal comments interpreting these passages—can we place compilation textbooks in tension with another concept of textbook that would be replaced in order to stabilize knowledge and standardize teaching and other practices.

Why does technical writing work to legitimate scientific knowledge in the 20th-century United States while other kinds of knowledge are marginalized or silent? The answers we find to this question will retell stories of struggles and tensions within our culture and within historic relations of knowledge and power. The histories we write in answer to this question will help us understand why we, as technical and professional communicators, value some types of knowledge, practice some types of communication, teach students some things and not others. These histories will help us understand how we arrived at our current situation, where we can go from here, and how we can get there. Technical writers and teachers concerned with multiculturalism, gender issues, conflict, and ethics can find roots of these current issues in the struggles that still reside in the language of our technical communications and standard teaching practices. By understanding the roots of these issues through systematic histories of our discourse, we can help to alloy the technical writer's scientific and liberal arts knowledge into new approaches to enduring human relations—approaches that may even seem unnatural or unscientific to our late-20th century sensibilities.

Because technical and professional communicators share a history as firmly rooted in the liberal arts as it is in the sciences, we occupy a unique position spanning these two disparate bodies of knowledge and approaches to making knowledge. We are in a position to combine the liberal arts and sciences in a system of teaching and communicating that draws on the strengths of both traditions, from scientific induction to rhetorical invention, from mechanical design to artistic visualization, from empirical observation to literary critique. Like our first- cousin composition studies, technical and professional communication share an openness to combining disparate research approaches to form an alloyed knowledge that is more robust than it could possibly be using one research approach alone. Teresa Kynell has observed that technical writing is "ultimately a recursive discipline that weds empiricism and rhetorical theory" (87). Using the strengths of both these science- and liberal arts-based research approaches, technical communicators are positioned to develop knowledge that compensates for weaknesses in any single way of knowing the world. Thus, in our professional practices we can embody the spirit of inclusion that underlies so many of our current professional concerns. These alloyed practices, based in both the sciences and arts, can create a genuinely revalued cultural currency for scientific knowledge made by technical writers.

TOWARD A HUMANISTIC TECHNICAL WRITING

In trying to conceptualize this alloyed technical writing, we could ask if and how a humanistic technical writing would differ from our current practices. Carolyn Miller asked this question in "A Humanistic Rationale for Technical Writing" nearly a decade ago and found that technical writing practices were based on a "pervasive positivist view of science" (610). Miller elaborated this positivist distinction between science and language: "In this view, human knowledge of which we may take science to be a model, is a matter of getting closer to the material things of reality and farther away from the confusing and untrustworthy imperfections of words and minds" (610). According to this distinction, technical writing "becomes the skill of subduing language so that it most accurately and directly transmits reality" (610). This view of technical writing as an invisible conduit transmitting reality through clear language has a venerable history in the writings of Bacon, the Royal Society, John Locke, T. A. Rickard, and any number of 20th-century technical writing textbook authors. It is a tradition within our discipline that is not easily transformed through humanistic influences.

Some practitioners have attempted to introduce humanistic knowledge into technical writing by calling on writing's rhetorical tradition. Most often, rhetoric is added as some instruction in audience awareness, as illustrated by Paul Anderson's Chapter 1 subheading in his textbook *Technical Writing: A Reader-Centered Approach*: "THE MAIN ADVICE OF THIS BOOK: THINK CONSTANTLY ABOUT YOUR READERS" (13). While Anderson's suggestion is sound rhetorical advice for any writer, it is traditional counsel for technical writers since Rickard reiterated this concise recommendation in *A Guide to Technical Writing* at the turn of the century: "Remember the reader" (110). It would seem that the invisible conduit paradigm of technical writing, on which Rickard clearly based his teachings, has traditionally accommodated at least a limited application of rhetorical knowledge. In order to alloy humanistic and scientific knowledge within technical information, therefore, we must look beyond these limited applications of rhetorical knowledge that are traditionally codified in our technical writing textbooks. These limited roles for rhetoric emphasize what is practical or efficient in the humanities within a knowledge/power system dominated by science, concurrently seeking to eliminate what is impractical or inefficient within this constrained system. Yet rhetoric and the humanities hold more promise—and even utility—than these limited roles allow.

At the end of the 20th century, many people in Western cultures are beginning to realize that positivist science—and its knowledge/power system—do not allow us to adequately address complex social issues that seem to defy remedy: environmental degradation, homelessness, teenage unwed parents, breakdown of family units, hate speech and actions, arms control, to name but a few of the more apparent issues. In a culture based on dominant scientific knowledge, these social problems are seen as problems for science (just as all problems are seen as nails to the person who only has a hammer). Yet scientists—as well as humanists, spiritualists, etc.—are beginning to question whether our scientific knowledge/power system can explain our social problems in ways that enable us to ameliorate them. Or does our scientific knowledge/power system disable us in the face of these complex social needs? Carolyn Miller argued in "Humanistic Rationale" that

> although our thinking about technical writing seems to be heavily indebted to the positivist view of science (and of rhetoric), this view is

no longer held by most philosophers of science or by most thoughtful scientists . . . a new epistemology, based on modern developments in cultural anthropology, cognitive psychology, and sociology, has challenged the positivist conceptions of knowledge. This new epistemology makes human knowledge thoroughly relative and science fundamentally rhetorical (615).

Here Miller found that philosophers and "thoughtful scientists" recognize that scientific knowledge is constructed by humans. In other words, the dominant scientific knowledge, on which our current knowledge/power system in based, is constructed by humans through a transformation of language into science. This observation begins to show us how we can see beyond our current scientific knowledge/power system to transform it into a system through which we can better address our complex social problems. Yet in Miller's explanation, this "new epistemology" would be based on "cultural anthropology, cognitive psychology, and sociology"—all scientized studies of human beings. By basing this "new epistemology" on humanistic studies that have been made into sciences, Miller stayed within the dominant scientific power/knowledge system instead of breaking an opening within its walls through which we might glimpse what lies beyond.

Current trends in technical communication research have responded to Miller's call for combining humanistic with scientific approaches to technical writing. In conceptualizing designs for technical writing programs, Billie Wahlstrom called for dexterity in various research methods: "[I]t doesn't matter what methodologies are privileged at technical communication research programs as long as students are required to develop an understanding and appreciation of multiple methods and the rigor with which these methods must be applied" (308). Wahlstrom cited the need for different methodological designs to address human aspects and social implications of research studies. Ann Hill Duin and Craig Hansen suggested issues for research in technical communication that begin to address these human aspects: contexts, power, authorship, connections. In discussing context, they argued for including the human dynamic in research designs: "Essentialist views of nonacademic writing that inappropriately freeze dynamic cultures, turning observation into rule, must be resisted" (11). In other words, research in technical writing can and should employ methods that get beyond a positivist view of the object of inquiry. These expanded methods can include critical research, as Carl Herndl advocated in discussing critical ethnography: "The arguments for redefining writing as social *praxis* may become part of our *habitus,* making it possible that the critique of contemporary ethnography will change our practice" (Herndl's italics, 31). The inclusion of humanistic, critical research can help researchers understand the social implications of technical writing practice. This expanded understanding of the implications of our practice can help us break openings in the wall of our power/knowledge system.

Making that opening in the wall of our dominant scientific knowledge/power system is an extremely difficult task, since our ways of thinking are so thoroughly shaped by this system and its legitimated practices. In the contemporary United States, each individual has internalized this system to differing degrees. How then, in light of our compliance with and adaptation to this system, can we reconceptualize what legitimate knowledge might be outside the dominant system? Of course, we cannot get entirely outside this system that has shaped us. But we can look to history to learn about times when what we would consider non-legitimate knowledge dominated and science was not legitimate, as in the 16th-century Germany of Agricola or the 17th-century England of Bacon. Or we can look to contemporary non-Western cultures to learn about alternate ways of conceptualizing the world and the place of humans within it. Closer to home, we can explore current non-dominant knowledges and practices in our own culture to find alternate and supplementary ways of approaching our social issues and transforming the knowledge/power system on which our technical writing practices are based.

Our current scientific knowledge/power system emphasizes individuals' logical and rational aspects, to the detriment of our emotional, intuitive, spiritual, and communal aspects (among other possibilities). It is these non-dominant human aspects that offer sites of alternative knowledge that can help technical writers reintroduce humanistic concerns to science and technology. Looking at emotions, for example, mechanical engineers who designed social systems in the late 19th century described how laborers did not *like* being systematized by the management systems those engineers designed. Similarly, industrial management researchers after World War II described how white-collar managerial workers did not *like* being controlled through systematized technical communications. In the 1980s, James Paradis, David Dobrin, and Richard Miller described how young research engineers at Exxon ITD did not *like* being controlled by their managers who reviewed reports the young engineers wrote. It seems that since the inception of systematized management, people have had negative emotional reactions to being systematized and controlled through technical communications. In reaction to this emotion, engineers and managers have sought to apply more science and systemizing to the problem, apparently not considering that a scientific approach alone might simply aggravate the negative emotions. If these emotions were not dismissed as non-legitimate or were not scientized into a sociological problem, we might be able to understand systematized management systems in a new light by exploring these emotions as legitimate human reactions pointing to limitations of management systems within human contexts. By putting humans—in all our complexity—at the center of our practices, we can begin to alloy a humanistic technical writing.

In working a transformation of technical writing—the currency of science—we cannot look to textbooks for guidance. After all, if textbooks are retrospective codifications of dominant professional practices, we can expect to find dominant practices from past decades in the textbooks that are profitably published and currently distributed. We cannot expect to learn non-standard practices—such as alloying scientific and liberal arts knowledge within technical communication—in textbooks. Instead, we can look to journal articles, current scholarly books, and debates in our discipline and other disciplines to discover ways in which we can modify our writing and teaching practices to include non-dominant knowledge. We can look to mass media to learn about cultural debates and issues that will modify our practices. But we must look within ourselves to remember what is human about technical writing in our scientific knowledge/power system and discover how we can reintroduce human complexity into that forgetful system.

Notes

INTRODUCTION

1. In discussing agents operating in scientific laboratories, Karin Knorr Cetina argued for a view of scientific knowledge-making in labs that includes an interrelated cultural context of scientists and instruments, as well as social, political, and scientific interests. Using this expanded view of laboratory culture, Knorr Cetina found that mundane lab practices became "epistemic devices in the production of knowledge" ("Couch" 119). Extending this notion, I would argue that technical writing also functions as one of these instruments or epistemic devices that scientists employ to make knowledge within cultural contexts.

2. Bruno Latour and Steve Woolgar also found that this combination of social and economic considerations was a more robust framework for understanding laboratory culture than a social framework alone (191–92).

3. Rickard's textbook, *A Guide to Technical Writing*, was originally published in 1908. A second edition was reprinted in 1910. I have used this second edition as my source for quotations throughout this history.

CHAPTER 1

1. For other examples of research exploring the cultural implications of technical writing practices, see Killingsworth and Palmer, "The Environmental Impact Statement and the Rhetoric of Democracy"; and three articles by Steven Katz, "Aristotle's Rhetoric, Hitler's Program, and the Ideological Problem of Praxis, Power and Professional Discourse," "The Ethic of Expediency," and "Narration, Technical Communication, and Culture."

2. See the following works discussing community affiliations: Carolyn Miller, "A Humanistic Rationale for Technical Writing"; Patricia Bizzell, "Cognition, Convention and Certainty"; Kenneth Bruffee, "Collaborative Learning and the 'Conversation of Mankind'"; Lester Faigley, "Nonacademic Writing"; and Cooper and Holzman *Writing as Social Action*.

3. See the following works discussing organizational or professional affiliations: Miller and Selzer, "Special Topics of Arguments in Engineering Reports"; Lucille McCarthy, "A Psychiatrist Uses *DSM-III*"; Amy Devitt, "Intertextuality in Tax Accounting"; Charlotte Thralls, "Rites and Ceremonials"; Carl Herndl, "Cultural Studies and Critical Science"; James Porter, "The Role of Law, Policy, and Ethics in Corporate Composing."

4. See the following works discussing relationships between writers and readers: Donald Rubin, "Introduction: Four Dimensions of Social Construction in Written Communication"; and Martin Nystrand, "A Social-Interactive Model of Writing."

5. See the following works attempting to synthesize a cognitive and social view of composing: Donald Rubin "Introduction: Four Dimensions of Social Construction in Written Communication"; and Nystrand, Greene, and Wiemelt, "Where Did Composition Studies Come From?"

6. See the following discussions of gender's influence on composing: Marilyn Cooper, "Women's Ways of Writing"; and Mary Lay, "Feminist Theory and the Redefinition of Technical Communication."

7. By "decontextualized" I mean a research approach that separates form from meaning and focuses primarily on the form, taking the meaning making as *a priori* and unproblematic. In *Fragments of Rationality*, Lester Faigley provides a more extended discussion of the split between form and meaning in interpretations of Saussure's work, explaining how U.S. composition researchers come from a linguistics tradition that focuses on the formal, structural aspects of Saussure's work, at the expense of the semantic or semiotic aspects.

8. In *Language and Symbolic Power*, Bourdieu introduces the idea of (mis)recognition to explain how people misrecognize states of being as natural instead of having been constructed through language at some point in time. This idea is useful for a cultural history of technical writing because it forms a foundation for questioning assumptions on which the practice of technical writing is based at a particular historical moment. Objections could be raised to this concept, however, on the basis that it implies there are objective "truths" that can be "recognized" through a proper philosophical framework—a position similar to the epistemology of positive science. Although such objections point to one limitation of this idea of (mis)recognition, I use Bourdieu's idea to highlight the position that some states of being, which we consider to be natural, are cultural and historical rather than being inevitable and universal.

9. For other examples of social constructionist research articles that misrecognize description of situated practice for prescriptions of generalized

practices for teaching, see Miller and Selzer, "Special Topics in Engineering Reports"; Gail Stygall, "Texts in Oral Context"; Amy Devitt, "Intertextuality in Tax Accounting"; Charlotte Thralls, "Rites and Ceremonials."

10. For a more complete exposition of this line of reasoning, see Longo, "Teaching Technical Communication with the Community Model."

11. See the following examples of using a limited view of culture in research design: Jennie Dautermann, "Negotiating Meaning in a Hospital Discourse Community"; Amy Devitt, "Intertextuality in Tax Accounting"; James Porter, "The Role of Law, Policy, and Ethics in Corporate Composing"; Graham Smart, "Genre as Community Invention"; Gail Stygall, "Texts in Oral Context"; and Charlotte Thralls, "Rites and Ceremonials."

12. See Michel de Certeau, *The Practice of Everyday Life*, pp. 65 ff. for a discussion of naive know-how.

13. See the following for discussions of multiculturalism in technical writing: Bell, Dillon, and Becker, "German Memo and Letter Style"; Kossek and Zonia, "The Effects of Race and Ethnicity on Perceptions of Human Resource Policies and Climate Regarding Diversity"; Nicholson et al., "United States versus Mexican Perceptions of the Impact of the North American Free Trade Agreement"; Beverly Sauer, "Communicating Risk in a Cross-Cultural Context"; and Michele Wender Zak, "'It's Like a Prison in There.'"

14. See the following for discussions of gender in technical communication: Boiarsky et al., "Men's and Women's Oral Communication in Technical/Scientific Fields"; Sam Dragga, "Women and the Profession of Technical Writing"; Griffeth et al.,"The Effects of Gender and Employee Classification Level on Communication-Related Outcomes"; and three articles by Mary Lay, "Feminist Theory and the Redefinition of Technical Communication," "Gender Studies: Implications for the Professional Communication Classroom," and "The Value of Gender Studies to Professional Communication Research."

15. See the following for discussions of conflict in technical communication: Rebecca Burnett, "Conflict in Collaborative Decision-Making"; Carl Herndl, "Teaching Discourse and Reproducing Culture"; and McCarthy and Gerring, "Revising Psychiatry's Charter Document DSM-IV."

16. See the following for discussions of ethics in technical communication: Stephen Doheny-Farina, "Research as Rhetoric"; Steven Katz, "The Ethic of Expediency"; and James Porter, "The Role of Law, Policy, and Ethics in Corporate Composing."

17. See the following discussions of community as a communication

model: Joseph Harris, "The Idea of Community in the Study of Writing"; Bruce Herzberg, "Rhetoric Unbound"; Thomas Kent, "On the Very Idea of a Discourse Community"; Carolyn Miller, "Rhetoric and Community"; Chantal Mouffe, "Democratic Citizenship and the Political Community"; John Trimbur, "Consensus and Difference in Collaborative Learning."

18. See the following for discussions of postmodernism in technical communication: Ben and Marthalee Barton, "Ideology and the Map"; Britt, Longo, and Woolever, "Extending the Boundaries of Rhetoric in Legal Writing Pedagogy"; and Richard Freed, "Postmodern Practice."

CHAPTER 2

1. For more complete discussions of Hermeticism, see Bruno, *The Expulsion of the Triumphant Beast*; Copenhaver, *Hermetica* (a translation of the *Corpus Hermetica*); Eamon, *Science and the Secrets of Nature*; Pumphrey, "The History of Science and the Renaissance Science of History"; Yates, *Giordano Bruno and the Hermetic Tradition*. For additional discussions of scientific knowledge in the early centuries A.D., see Bowen, *A History of Western Education* Volume I; Stahl, *Roman Science*; Wagner, *The Seven Liberal Arts in the Middle Ages*.

2. See William Eamon, Chapter 2 "Knowledge and Power" and Stuart Clark, "The Rational Witchfinder" for more complete discussions of the role of magic in relation to science and technology in medieval and Renaissance European cultures.

3. Although Agricola was one of many handbook writers, his approach to choosing which information to include in the handbook was unlike other writers. By including Hermetic secrets and arguing for their utility while distancing himself from their occult nature, Agricola distinguished his handbooks by their comprehensive treatments of what had been occult or blasphemous knowledge.

4. Roger Bacon is credited with using an experimental approach to study natural phenomena in the mid-13th century. For example, Stephen Pumphrey noted that both Roger Bacon and Theodoric of Freibourg conducted "optical researches" (53), although experimental practices were not common in 13th-century science. J.G. Crowther also noted that "Bacon made considerable contributions to optics" (151), as well as other fields, such as the study of gunpowder and geography. He quoted Bacon as writing that he had "'learned more useful and excellent things without comparison from very plain people unknown to fame in letters, than from all [his] famous teachers'" (152). In his experimental studies of natural phenomena and his emphasis on learning from sources out-

side adademic authority, Roger Bacon presents a figure of modern science. His adherence to Christian theological doctrines and their social implications, however, implies that Roger Bacon studied experimental science without advocating social change through science.

William of Ockham sought to differentiate religion from science in the 14th century. Ockham set out a proto-evolutionary thesis when he wrote (as quoted in Crowther): "'[N]o human institution is absolute or final, and neither Pope nor Emperor can claim exemption from the general law of progress and adaptation'" (159). Crowther found Ockham's view to be "inspired by his part in the class conflict between Church and state" (159). Seth, too, noted that Ockham struggled against "papal authority" (15). Peter Dear, however, argued that although Ockham's philosophical opinions resulted in "political and theological difficulties" (120), they were tolerated by the Church. Evidently, the 14th-century Church did not view Ockham's religious and social opinions as threatening to Church authority.

Roger Bacon and William of Ockham each addressed a part of what Francis Bacon would combine in his program for a public science: Roger Bacon focused on experimental methods while Ockham focused on a social program that questioned *a priori* authority, such as that embodied by the Church. I have chosen Francis Bacon as a starting point for this study because his works combine both these scientific and social aspects of experimental science and anti-scholastic philosophy.

5. William Eamon discussed a long line of writers who wrote against the academic, scholastic tradition of logical speculation. These writers compiled books of secrets gained from experience and observation of animal behaviors. For example, Leonardo Fioravanti (1518–1588) "extolled practical experience, common sense and sound judgment" over formal education (183). Giambattista Della Porta (1535?–1615) "dedicated his life to establishing natural magic as a legitimate empirical science" (196). Numerous other 16th-century authors advocated medical remedies based on the "raw empiricism of the 'hand-in-the-wound' school" (260) rather than those developed in medical schools. In *Ancients and Moderns*, Richard Jones also noted that Robert Norman's *The Newe Attractius* (1581) and William Gilbert's *De Magnete* (1600) were English forebears of Bacon's anti-scholasticism.

6. I use this reductionist identification of "the Christian Church" to simplify this brief discussion of Francis Bacon's intellectual context. As Peter Dear pointed out in "The Church and the New Philosophy," though, the social milieu of Renaissance Europe was much more complex than this phrase "the authority of the Christian Church" indicates. For example, Catholic and Protestant churches interacted in Renaissance Europe, and the status of the Church in England was particularly blurred from the reign of Henry VIII until William

and Mary took the throne in 1689 to affirm that country's anti-Catholic rule. Even within the Catholic Church, practices were not as homogeneous as might appear at first look. Many Jesuit scholars embraced Copernican astronomy even though it was officially banned by the Catholic Church. (See J. A. Bennett, "The Challenge of Practical Mathematics.") Books on the *Index* were not always destroyed or unread, especially in rural areas where lax Church officials may not enforce the Vatican's rules. (See Paolo Rossi, "Society, Culture and the Dissemination of Learning.")

7. For a discussion of this Renaissance view of the decline of the human condition, see Richard Jones, *Ancients and Moderns*, Chapter 2 "The Decay of Nature."

8. According to J. G. Crowther, Francis Bacon's father, Sir Nicholas Bacon, was "one of Elizabeth's great statesmen" (251), serving as Lord Keeper of the Great Seal from the beginning of the Queen's reign in 1558. Julian Martin noted that Bacon's uncles—Sir Thomas Gresham, Sir William Cecil (Lord Burleigh), Sir Henry Killegrew, and Sir Thomas Hoby—"all entered the service of the Tudor dynasty during the 1530s and 1540s, and they served it throughout their lives" (106). Of these uncles, James Creighton considered Lord Burleigh, who served as Lord Treasurer, to be "perhaps the most prominent statesman of the time" (iv). Farrington called Sir Nicolas Bacon and Lord Burleigh the "twin pillars of the realm" (12). Farrington also argued that from boyhood, Francis Bacon intended to implement his philosophy for a public science throught administrative means: " . . . at first he thought of introducing his reform administratively. Administrative action was a natural line, considering his family position, for his ambition to take" (12). Although Francis Bacon "never received an important appointment from Elizabeth" (Burtt 3), Robert Faulkner noted, "Under James I, however, Bacon rose spectacularly to become the second or third most powerful personage in the kingdom. He was appointed solicitor-general (1607), attorney-general (1613), and then, in 1618, Lord Chancellor (the chief law officer) for life" (4). Bacon fell from power in 1621 as spectacularly as he had risen to it, after pleading guilty to accepting bribes from litigants—a common practice even when cases were pending. Creighton quoted Bacon as saying at the time, "I was the justest judge that was in England these fifty years; but it was the justest censure that was in Parliament these two hundred years" (vi).

CHAPTER 3

1. For other discussions of the influences of Francis Bacon on the use of plain language in science, see John Briggs, "Francis Bacon and the Rhetoric of

Nature"; Morris Croll, "Attic Prose"; Richard Jones, "Science and English Prose Style in the Third Quarter of the Seventeenth Century"; George Williamson, *The Senecan Amble*; and two articles by James Zappen, "Francis Bacon and the Historiography of Scientific Rhetoric" and "Francis Bacon on Democratic Science and Plain Prose."

2. Huxley's son Leonard provided details of the publication of this monograph in *Life and Letters of Thomas Henry Huxley*, pp. 531 ff.

3. For other renditions of this quotation of Bacon, see the following Huxley lectures: "Technical Education" (1877) and "Address on behalf of the National Association for the Promotion of Technical Education" (1887). Another quote from Bacon can be found at the end of "On Science and Art in Relation to Education " (1882).

4. See Leonard Huxley's *Life and Letters* pp. 520 ff. and 535 for a further discussion of T. H. Huxley's opinions on Francis Bacon.

CHAPTER 5

1. For other expressions of engineers' concern with moral and material progress in the late 19th and early 20th century, see Harrington Emerson's *Efficiency as a Basis for Operation and Wages* and "Efficiency in the Manufacture of Railway Transportation," H. L. Gantt's serialized articles on "The Compensation of Workmen," the discussion following F. A. Halsey's "The Premium Plan of Paying for Labor," Francis Lyster Jandron's "Efficiency and the Railway Wage Problem," the discussion following Frank Richard's "Is Anything the Matter with Piece Rate Work?," Frederick W. Taylor's "A Piece-Rate System," and the discussion following Henry Towne's "Gain-Sharing."

2. Henry Towne was president of the American Society of Mechanical Engineers in 1888 when he presented this paper on gain-sharing. His position within the Society indicates the importance of the pay system question for Society members during the late-19th century.

3. For a complete discussion of the "labor problem" at the Ford Motor Company, see Mark Rupert's *Producing Hegemony*. Also see other discussions of Fordism in Hounshell's Chapter 6, "The Ford Motor Company and the Rise of Mass Production in America" in *From the American System to Mass Production*; Hughes' Chapter 5, "The System Must Come First" in *American Genesis*; Smith's Chapter 1, "Fordism: Mass Production and Total Control" in *Making the Modern*.

4. Scientific management developer Frederick W. Taylor was an alumnus of Stevens Institute of Technology, having earned a Mechanical Engineering (M.E.) degree from that Hoboken, New Jersey, school in 1883. See Wrege and Greenwood, pp. 34–35.

CHAPTER 6

1. For discussions of the influence of Taylor's system of scientific management on businesses in the United States, see Taylor's "Testimony Before the Special House Committee" in *Scientific Management*, Navin's Chapter 15, Section 3 "Scientific Management" in *The Whitin Machine Works Since 1831*; The Taylor Society's Chapter 3, "The Influence of Scientific Management" in *Scientific Management in American Industry*; and Urwick and Brech's "Scientific Management and Society" in *The Making of Scientific Management*, Volume I.

2. During the 14th century, the term "liberal arts" was used to distinguish skills appropriate to independent or free (liberal) men from those "mechanical" skills appropriate to a lower class of indentured men. Even from this early time, there was a tension within the word liberal between its meaning as "generous" (gifts given from the free classes to the indentured classes) and its meaning as "unrestrained" (excesses engaged in by the free classes) (Williams 179). In the 16th century, the word "artist" began to be used to describe a person skilled in the liberal arts of grammar, logic, rhetoric, arithmetic, geometry, music, and astronomy.

In the 17th century, the term "science" was still interchangeable with the term "art" to describe a particular body of knowledge or skill. By the early 18th century, the term "science" was distinguished from "art" in that science dealt with the experimental and external while art dealt with the experiential and internal. Science became the methodical and theoretical study of nature; art became the methodical and theoretical study of experiences (religious, metaphysical, social, political, feeling, inner life). Art became not-science (Williams 277–78), at which point scientific knowledge can be said to have conquered artistic knowledge because science had become the standard against which all other knowledges were measured.

In 1840, William Whewell distinguished between the scientist and artist: "We need very much a name to describe a cultivator of science in general. I should incline to call him a Scientist. Thus we might say, that as an Artist is a Musician, Painter, or Poet, a Scientist is a Mathematician, Physicist, or Naturalist" (cxiii). James Paradis described an early discussion of methodological differences between scientists and artists as set out by Romantic artist David Scott, also in 1840: "'Leonardo was mentally a seeker after truth—a scientist.

Correggio was an asserter of truth—an artist'" (*Huxley* 12). Science became the model of an external object of study and a neutral methodical observer pursuing the truth, fact, reason.

Until the 18th century, most sciences were seen to be arts. But by the mid-19th century, science was distinguished from art as requiring different skills and efforts, and having different methods and purposes. The definition of scientist allowed for the specialization of the artist by the late 19th century, and the distinction between the fine arts (painting, drawing, and sculpture) and the liberal arts.

CHAPTER 7

1. Although the ranks of non-engineer technical writers became larger and more influential after World War II, many engineers continued to communicate their findings in the earlier tradition of engineer/writers. This co-existence of engineers and non-engineers making technical knowledge continues today.

2. Niles, Niles, and Stephens include a similar explanation of the growth of clerical workers in *The Office Supervisor* (1959): "From decade to decade the office has been growing in importance in American life until in 1958 there were nine million clerical employees—one in seven of all persons gainfully employed. . . . A number of major reasons may be cited for the growing volume of office workers. 1. There has been a great growth in the number and size of office-type institutions . . . 2. There has been an increasing necessity for records and reports in all organizations . . . 3. Not least are the reporting and audit requirements imposed by federal, state, and local government . . . 4. Management itself requires more and more data for analysis . . . 5. The growth in size of organization has required more paper work" (10).

3. Taylor's scientific management system can be seen as late as 1980 in a work by Nolan, Young, and DiSylvester entitled *Improving Productivity Through Advanced Office Controls*. In this late version of scientific management, the Advanced Office Controls consist of work simplification, time study, standards setting, and work measurement, in much the same form they were presented at the turn of the century.

4. Hotchkiss is listed as the head of the Department of Advertising and Marketing at New York University.

5. The editor's biography of James Thomson Shotwell in this volume included the following information about Shotwell's career: "During the World War [I] he was Chairman of the National Board for Historical Service; and after

the Armistice, a member of the Preparatory Commission for the Peace Conference, at which he served as Chief of the Division of History. . . . Since 1924 he has been Director of the Division of Economics and History of the Carnegie Endowment for International Peace. He is the author and editor of many volumes dealing with the origins and results of the World War, with labor problems, and with international relations" (201). Shotwell's work in history and international relations suggests he had an ongoing commitment to what would traditionally be considered humanistic pursuits. Yet in "Mechanism and Culture" his condemnation of artists and literary writers is as forceful as his praise of mechanical practicality.

6. In their textbook *Technical Writing* (1960), for example, Wellborn, Green, and Nall included description under the heading of "Definition and Description" (xi), even though definition had traditionally been one of the subcategories of exposition. Technical writing textbook author George Harwell went even further in subsuming explication, description, definition, and narration all under the heading of "Methods of Exposition" (vii) in *Technical Communication* (1969).

References

Agricola, Georgius. *De Re Metallica*. Trans. Herbert C. Hoover and Lou Henry Hoover. 1556. New York: Dover Publications, 1950.

Alexander, Peter. "Solidity and Elasticity in the Seventeenth Century." *Locke's Philosophy: Content and Context*. Ed. G. A. J. Rogers. Oxford: Clarendon Press, 1994. 143–64.

Anderson, Paul V. *Technical Writing: A Reader-Centered Approach*. 2nd ed. Fort Worth: Harcourt Brace Jovanovich, 1991.

Andrade, E. N. da C. *A Brief History of the Royal Society*. London: The Royal Society, 1960.

Bacon, Francis. *The Great Instauration and New Atlantis*. Ed. J. Weinberger. Arlington Heights: AHM Publishing Corp., 1980.

———. *Novum Organum with Other Parts of The Great Instauration*. Trans. and ed. Peter Urbach and John Gibson. 1620. Chicago: Open Court, 1994.

Baldwin, Charles Sears. *Composition Oral and Written*. New York: Longmans, Green and Co., 1909.

Barke, Richard. *Science, Technology, and Public Policy*. Washington, D.C.: The Congressional Quarterly Press, 1986.

Barthes, Roland. "The Great Family of Man." *Mythologies*. Trans. Annette Lavers. New York: The Noonday Press, 1972.

Barton, Ben F., and Marthalee S. Barton. "Ideology and the Map: Toward a Postmodern Visual Design Practice." *Professional Communication: The Social Perspective*. Ed. Nancy Roundy Blyler and Charlotte Thralls. Newbury Park: Sage, 1993. 49–78.

Bell, Arthur H., W. Tracy Dillon, and Harald Becker. "German Memo and Letter Style." *Journal of Business and Technical Communication* 9.2 (April 1995): 219–27.

Benjamin, Walter. "The Work of Art in the Age of Mechanical Reproduction." *Illuminations*. Trans. Harry Zohn. Ed. Hannah Arendt. New York: Schocken Books, 1968. 217–52.

———. "Theses on the Philosophy of History." *Illuminations*. Trans. Harry Zohn. Ed. Hannah Arendt. New York: Schocken Books, 1968. 253–64.

Bennett, J. A. "The Challenge of Practical Mathematics." *Science, Culture and Popular Belief in Renaissance Europe.* Ed. Stephen Pumfrey, Paolo L. Rossi, and Maurice Slawinski. Manchester: Manchester University Press, 1991. 176–90.

Berk, Gerald. *Alternative Tracks: The Constitution of American Industrial Order, 1865–1917.* Baltimore: The Johns Hopkins University Press, 1994.

Berkenkotter, Carol, Thomas N. Huckin, and John Ackerman. "Social Context and Socially Constructed Texts: The Initiation of a Graduate Student into a Writing Research Community." *Textual Dynamics of the Professions: Historical and Contemporary Studies of Writing in Professional Communities.* Ed. Charles Bazerman and James Paradis. Madison: University of Wisconsin Press, 1991. 191–215.

Bibby, Cyril. *Scientist Extraordinary: The Life and Scientific Work of Thomas Henry Huxley 1825–1895.* Oxford: Pergamon Press, 1972.

Blinderman, Charles S. "T. H. Huxley's Theory of Aesthetics: Unity in Diversity." *The Journal of Aesthetics and Art Criticism* 21 (1962): 49–55.

Bizzell, Patricia. "Cognition, Convention, and Certainty: What We Need to Know about Writing." *Pre/Text* 3.3 (1982): 213–43.

Bizzell, Patricia, and Bruce Herzberg, ed. *The Rhetorical Tradition: Readings from Classical Times to the Present.* Boston: Bedford Books, 1990.

Boiarsky, Carolyn, et al. "Men's and Women's Oral Communication in Technical/Scientific Fields: Results of a Study." *Technical Communication* 42.3 (August 1995): 451–59.

Boisjoly, Roger, engineer with Morton Thiokol. *Company Loyalty and Whistle Blowing: Ethical Decisions and the Space Shuttle Disaster.* A talk at Massachusetts Institute of Technology. Videocassette. January 7, 1987.

Bourdieu, Pierre. *Language and Symbolic Power.* Trans. Gino Raymond and Matthew Adamson. Cambridge: Harvard University Press, 1991.

Bowen, James. *A History of Western Education.* 3 vols. New York: St. Martin's Press, 1972–1981.

Briggs, John C. *Francis Bacon and the Rhetoric of Nature.* Cambridge: Harvard University Press, 1989.

Britt, Elizabeth, Bernadette Longo, and Kristin Woolever. "Extending the Boundaries of Rhetoric in Legal Writing Pedagogy." *Journal of Business and Technical Communication* 10.2 (April 1996): 213–38.

Bruffee, Kenneth. "Collaborative Learning and the 'Conversation of Mankind.'" *College English* 46 (1984): 635–52.

Bruno, Giordano. *The Expulsion of the Triumphant Beast.* Trans. and ed. Arthur D. Imerti. 1585. New Brunswick: Rutgers University Press, 1964.

Bryan, William Jennings. "The Cross of Gold Speech." *Great American Speeches*. Ed. Gregory R. Suriano. New York: Gramercy Books, 1993. 114–18.

Burnett, Rebecca. "Conflict in Collaborative Decision-Making." *Professional Communication: The Social Perspective*. Ed. Nancy Roundy Blyler and Charlotte Thralls. Newbury Park: Sage, 1993. 144–62.

———. *Technical Communication*. 2nd ed. Belmont: Wadsworth Publishing, 1990.

Burtt, Edwin A. *The English Philosophers from Bacon to Mill*. New York: Modern Library, 1939.

Bush, Vannevar. *Science, the Endless Frontier: A Report to the President on a Program for Postwar Scientific Research*. 1945. Washington, D.C.: National Science Foundation, 1960.

Carey, Charles M. *Business Writing*. Notre Dame: University of Notre Dame, 1949.

Carpenter, Stanley R. "Instrumentalists and Expressivists: Ambiguous Links between Technology and Democracy." *Democracy in a Technological Society*. Ed. Langdon Winner. Dordrecht: Kluwer Academic Publishers, 1992. 161–74.

Casey, Daniel Vincent. "Putting Standards into Practice." *How Scientific Management Is Applied*. 1911. Easton: Hive Publishing Company, 1974.

Chandler, Alfred D., Jr. *Strategy and Structure: Chapters in the History of the Industrial Enterprise*. Cambridge: The MIT Press, 1962.

Chappell, Vere. *The Cambridge Companion to Locke*. Cambridge: Cambridge University Press, 1994.

Church, A. Hamilton. "The Meaning of Commercial Organisation." *The Engineering Magazine* 20.3 (December 1900): 391–98.

———. *The Proper Distribution of Expense Burden*. New York: *The Engineering Magazine*, 1908.

Connors, Robert. "The Rhetoric of Mechanical Correctness." *Only Connect*. Ed. Thomas Newkirk. Upper Montclair, New Jersey: Cook Publishers, Inc., 1986. 27–58.

———. "The Rise and Fall of the Modes of Discourse." *The Writing Teacher's Sourcebook*. 2nd ed. Ed. Gary Tate and Edward P. J. Corbett. New York: Oxford University Press, 1988. 24–33.

———. "The Rise of Technical Writing Instruction in America." *Journal of Technical Writing and Communication* 12.4 (1982): 329–52.

Cooper, Marilyn. "Women's Ways of Writing." *Writing as Social Action*. Ed. Marilyn Cooper and Michael Holzman. Portsmouth: Heinemann, 1989. 141–56.

Cooper, Marilyn, and Michael Holzman. *Writing as Social Action*. Portsmouth: Heinemann, 1989.

Copenhaver, Brian. *Hermetica*. Cambridge: Cambridge University Press, 1992.

Cortada, James W. *Before the Computer: IBM, NCR, Burroughs, and Remington Rand and the Industry They Created, 1865–1956.* Princeton: Princeton University Press, 1993.

Creighton, James Edward. "Special Introduction." *Advancement of Learning and Novum Organum.* Francis Bacon. New York: P.F. Collier & Son, 1900.

Croll, Morris W. "Attic Prose: Lipsius, Montaigne, Bacon." *"Attic" and Baroque Prose Style: The Anti-Ciceronian Movement: Essays by Morris W. Croll.* Ed. J. Max Patrick and Robert O. Evans, with John M. Wallace. 1966. Princeton: Princeton Paperbacks-Princeton University Press, 1969. 167–202.

Crouch, W. George, and Robert L. Zetler. *A Guide to Technical Writing.* New York: The Ronald Press Company, 1948.

Crowther, J. G. *The Social Relations of Science.* Rev. ed. Chester Springs: Dufour, 1967.

Dautermann, Jennie. "Negotiating Meaning in a Hospital Discourse Community." *Writing in the Workplace: New Research Perspectives.* Ed. Rachel Spilka. Carbondale: Southern Illinois University Press, 1993. 98–110.

Dear, Peter. "The Church and the New Philosophy." *Science, Culture and Popular Belief in Renaissance Europe.* Ed. Stephen Pumfrey, Paolo L. Rossi, and Maurice Slawinski. Manchester: Manchester University Press, 1991. 119–42.

de Certeau, Michel. *The Practice of Everyday Life.* Trans. Steven Rendall. 1984. Berkeley: University of California Press, 1988.

———. *The Writing of History.* Trans. Tom Conley. 1974. New York: Columbia University Press, 1988.

Derrida, Jacques. "Plato's Pharmacy." *Disseminations.* Trans. Barbara Johnson. 1972. Chicago: University of Chicago Press, 1981. 61–172.

Devitt, Amy J. "Intertextuality in Tax Accounting: Generic, Referential, and Functional." *Textual Dynamics of the Professions: Historical and Contemporary Studies of Writing in Professional Communities.* Ed. Charles Bazerman and James Paradis. Madison: University of Wisconsin Press, 1991. 336–57.

Doheny-Farina, Stephen. "Research as Rhetoric: Confronting the Methodological and Ethical Problems of Research on Writing in Nonacademic Settings." *Writing in the Workplace: New Research Perspectives.* Ed. Rachel Spilka. Carbondale: Southern Illinois University Press: 1993. 253–67.

———. *Rhetoric, Innovation, Technology: Case Studies of Technical Communication in Technology Transfers.* Cambridge: The MIT Press, 1992.

Donald, W. J., ed. *Handbook of Business Administration.* 1st ed. New York: McGraw-Hill, 1931.

Dragga, Sam. "Women and the Profession of Technical Writing: Social and Economic Influences and Implications." *Journal of Business and Technical Communication* 7.3 (July 1993): 312–21.

Driskill, Linda. *Business and Managerial Communication: New Perspectives*. Fort Worth: Harcourt Brace Jovanovich, 1992.

Duin, Ann Hill, and Craig J. Hansen. "Setting a Sociotechnological Agenda in Nonacademic Writing." *Nonacademic Writing: Social Theory and Technology*. Ed. Ann Hill Duin and Craig J. Hansen. Mahwah: Lawrence Erlbaum Associates, 1996. 1–16.

Eamon, William. *Science and the Secrets of Nature: Books of Secrets in Medieval and Early Modern Culture*. Princeton: Princeton University Press, 1994.

Ellul, Jaçques. *The Technological Bluff*. Trans. Geoffrey W. Bromiley. Grand Rapids: William B. Eerdmans Publishing Company, 1990.

Emerson, Harrington. *Efficiency as a Basis for Operation and Wages*. 1908. Easton: Hive Publishing Company, 1976.

———. "Efficiency in the Manufacture of Railway Transportation: II. The Influence of the Personality of the Railroad Executive." *The Engineering Magazine* 44.4 (January 1913): 481–86.

———. *The Twelve Principles of Efficiency*. 1912. Easton: Hive Publishing Company, 1976.

Faigley, Lester. *Fragments of Rationality: Postmodernity and the Subject of Composition*. Pittsburgh: University of Pittsburgh Press, 1992.

———. "Nonacademic Writing: The Social Perspective." *Writing in Nonacademic Settings*. Ed. Lee Odell and Dixie Goswami. New York: Guilford, 1985. 231–48.

Farrington, Benjamin. *The Philosophy of Francis Bacon: An Essay on its Development from 1603–1609*. Chicago: The University of Chicago Press, 1966.

Faulkner, Robert K. *Francis Bacon and the Project of Progress*. Lanham: Rowman & Littlefield Publishers, 1993.

Florman, Samuel C. *The Existential Pleasures of Engineering*. New York: St. Martin's Press, 1976.

Folsom, M. B. "The Field of Office Management." *Handbook of Business Administration*. 1st ed. Ed. W. J. Donald. New York: McGraw-Hill, 1931. 743–48.

Foucault, Michel. *Archaeology of Knowledge*. Trans. A. M. Sheridan Smith. 1969. New York: Barnes & Noble, 1972.

———. *The Birth of the Clinic*. Trans. A. M. Sheridan. 1963. London: Routledge, 1973.

———. *Discipline and Punish: The Birth of the Prison.* 1975. Trans. Alan Sheridan. New York: Vintage Books, 1979.

———. *The Order of Things: An Archaeology of the Human Sciences.* 1966. New York: Vintage Books, 1994.

———. *Power/Knowledge: Selected Interviews and Other Writings 1972–77.* Trans. Colin Gordon et al. Ed. Colin Gordon. New York: Pantheon, 1980.

Francis, W. Nelson. *The English Language, An Introduction: Background for Writing.* New York: W. W. Norton & Company, 1965.

Freed, Richard C. "Postmodern Practice: Perspectives and Prospects." *Professional Communication: The Social Perspective.* Ed. Nancy Roundy Blyler and Charlotte Thralls. Newbury Park: Sage, 1993. 196–214.

Gantt, H. L. "The Compensation of Workmen." *The Engineering Magazine* 38.5 (February 1910): 653–62.

———. "The Compensation of Workmen and Efficiency of Operation: III. Task and Bonus." *The Engineering Magazine* 39.1 (April 1910): 17–23.

Genung, John Franklin. *The Practical Elements of Rhetoric with Illustrative Examples.* Boston: Ginn & Company, 1886.

———. *The Working Principles of Rhetoric.* Boston: Ginn & Company, 1901.

Giard, Luce. "Remapping Knowledge, Reshaping Institutions." *Science, Culture and Popular Belief in Renaissance Europe.* Ed. Stephen Pumfrey, Paolo L. Rossi, and Maurice Slawinski. Manchester: Manchester University Press, 1991. 19–47.

Gilbert, G. Nigel, and Michael Mulkay. *Opening Pandora's Box: A Sociological Analysis of Scientists' Discourse.* Cambridge: Cambridge University Press, 1984.

Gould, Jay R. *Opportunities in Technical Writing.* 1st ed. New York: Vocational Guidance Manuals, 1964.

Griffeth, Roger W., et al. "The Effects of Gender and Employee Classification Level on Communication-Related Outcomes: A Test of Structuralist and Socialization Hypotheses." *Journal of Business and Technical Communication* 8.3 (July 1994): 299–318.

Grillo, Elmer V. *Control Techniques for Office Efficiency.* New York: McGraw-Hill, 1963.

Grillo, Elmer V., and Charles J. Berg, Jr. *Work Measurement in the Office: A Guide to Office Cost Control.* New York: McGraw-Hill, 1959.

Halloran, Michael. "From Rhetoric to Composition: The Teaching of Writing in America to 1900." *A Short History of Writing Instruction: From Ancient Greece to Twentieth-Century America.* Ed. James J. Murphy. Davis: Hermagoras Press, 1990. 151–82.

Halsey, F. A. "The Premium Plan of Paying for Labor." *Transactions of the American Society of Mechanical Engineers* 12 (1891): 755–80.

Harrington, John Lyle. "The Value of English to the Technical Man." *Engineering Education: Essays for English*. 2nd ed. Ed. Ray Palmer Baker. New York: John Wiley & Sons, 1928. 87–103.

Harris, Joseph. "The Idea of Community in the Study of Writing." *College Composition and Communication* 40 (1989): 11–22.

Harwell, George C. *Technical Communication*. Toronto: The Macmillan Company, 1969.

Hatch, Edwin. *The Influence of Greek Ideas on Christianity*. 1888. New York: Harper & Row, 1957.

Herndl, Carl G. "Cultural Studies and Critical Science." *Understanding Scientific Prose*. Ed. Jack Selzer. Madison: University of Wisconsin Press, 1993a. 61–81.

———. "Teaching Discourse and Reproducing Culture: A Critique of Research and Pedagogy in Professional and Non-Academic Writing." *College Composition and Communication* 44 (1993b): 349–63.

———. "The Transformation of Critical Ethnography into Pedagogy, or the Vicissitudes of Traveling Theory." *Nonacademic Writing: Social Theory and Technology*. Ed. Ann Hill Duin and Craig J. Hansen. Mahwah: Lawrence Erlbaum Associates, 1996. 17–34.

Herndl, Carl, Barbara Fennell, and Carolyn Miller. "Understanding Failures in Organizational Discourse: The Accident at Three Mile Island and the Shuttle Challenger Disaster." *Textual Dynamics of the Professions: Historical and Contemporary Studies of Writing in Professional Communities*. Ed. Charles Bazerman and James Paradis. Madison: University of Wisconsin Press, 1991. 279–305.

Herzberg, Bruce. "Michel Foucault's Rhetorical Theory." *Contending with Words: Composition and Rhetoric in a Postmodern Age*. Ed. Patricia Harkin and John Schilb. New York: Modern Language Association, 1991. 69–81.

———. "Rhetoric Unbound: Discourse, Community, and Knowledge." *Professional Communication: The Social Perspective*. Ed. Nancy Roundy Blyler and Charlotte Thralls. Newbury Park: Sage, 1993. 35–48.

Hjortshoj, Keith. "The Marginality of the Left-Hand Castes (A Parable for Writing Teachers)." *College Composition and Communication* 46.4 (December 1995): 491–505.

Hotchkiss, George Burton, and Edward Jones Kilduff. *Handbook of Business English*. Rev. ed. New York: Harper & Brothers, 1917.

Hounshell, David A. *From the American System to Mass Production 1800–1932: The Development of Manufacturing Technology in the United States*. Baltimore: The Johns Hopkins University Press, 1984.

Houp, Kenneth W., and Lt. Colonel Thomas E. Pearsall. *Reporting Technical Information*. Beverly Hills: The Glencoe Press, 1968.

Houp, Kenneth W., and Thomas E. Pearsall. *Reporting Technical Information*. 3rd ed. Encino: Glencoe Publishing Co., 1977.

Hughes, Thomas P. *American Genesis: A Century of Invention and Technological Enthusiasm 1870–1970*. New York: Penguin Books, 1989.

Humphreys, Alexander C. "The Present Opportunities and Consequent Responsibilities of the Engineer." *Transactions of the American Society of Mechanical Engineers* 34 (1912): 607–37.

Huxley, Leonard. *Life and Letters of Thomas Henry Huxley*. 2 vols. New York: D. Appleton and Company, 1916.

Huxley, Thomas H. *Hume: With Helps to the Study of Berkeley*. New York: D. Appleton & Company, 1896.

———. *Science and Culture and Other Essays*. New York: D. Appleton & Company, 1882.

———. *Science and Education: Essays*. New York: D. Appleton & Company, 1896.

Jackson, Dugald C. "The Origins of Engineering." *Science* 78:2035 (December 29, 1933): 589–96.

Jakubec, Jan. *Johannes Amos Comenius*. New York: Arno Press & The New York Times, 1971.

Jandron, Francis Lyster. "Efficiency and the Railway Wage Problem." *The Engineering Magazine* 44.2 (November 1912): 241–47.

Jastram, Roy W. *The Golden Constant*. New York: John Wiley & Sons, 1977.

Jelinek, Mariann. "Toward Systematic Management: Alexander Hamilton Church." *Business History Review* 54:1 (Spring 1980): 63–79.

Jenkins, John J. *Understanding Locke: An Introduction to Philosophy through John Locke's Essay*. Edinburgh: Edinburgh University Press, 1983.

Johnson, John Butler. "Two Kinds of Education for Engineers." *Engineering Education: Essays for English*. 2nd ed. Ed. Ray Palmer Baker. New York: John Wiley & Sons, 1928. 67–83.

Johnson, Nan. *Nineteenth-Century Rhetoric in North America*. Carbondale: Southern Illinois University Press, 1991.

Jones, Richard Foster. *Ancients and Moderns: A Study of the Rise of the Scientific Movement in Seventeenth-Century England*. 2nd ed. St. Louis: Washington University Press, 1961.

———. "Science and English Prose Style in the Third Quarter of the Seventeenth Century." *The Seventeenth Century: Studies in the History of English*

Thought and Literature from Bacon to Pope. Stanford: Stanford University Press, 1951. 75–110.

Katz, Steven. Aristotle's Rhetoric, Hitler's Program, and the Ideological Problem of Praxis, Power and Professional Discourse." *Journal of Business and Technical Communication* 7.1 (January 1993): 37–62.

———. "The Ethic of Expediency: Classical Rhetoric, Technology, and the Holocaust." *College English* 54 (1992a): 255–75.

Keller, Evelyn Fox. *Reflections on Gender and Science.* New Haven: Yale University Press, 1985.

Kent, Thomas. "On the Very Idea of a Discourse Community." *College Composition and Communication* 42 (1991): 425–46.

Kitzhaber, Albert R. *Rhetoric in American Colleges, 1850–1900.* Dallas: Southern Methodist University Press, 1990.

Kleimann, Susan. "The Reciprocal Relationship of Workplace Culture and Review." *Writing in the Workplace: New Research Perspectives.* Ed. Rachel Spilka. Carbondale: Southern Illinois University Press: 1993. 56–70.

Kline, Ronald. "Construing 'Technology' as 'Applied Science.'" *Isis* 86 (June 1995): 194–221.

Knorr Cetina, Karin. "Producing and Reproducing Knowledge: Descriptive or Constructive? *Sociology of Science* 16 (1977): 669–96.

———. "The Couch, the Cathedral, and the Laboratory: On the Relationship between Experiment and Laboratory in Science. *Science as Practice and Culture.* Ed. Andrew Pickering. Chicago: University of Chicago Press, 1992. 113–38.

Kossek, Ellen Ernst, and Susan C. Zonia. "The Effects of Race and Ethnicity on Perceptions of Human Resource Policies and Climate Regarding Diversity." *Journal of Business and Technical Communication* 8.3 (July 1994): 319–34.

Kuhn, Thomas S. *The Structure of Scientific Revolutions.* 2nd ed. Chicago: University of Chicago Press, 1970.

Kynell, Teresa C. *Writing in a Milieu of Utility: The Move to Technical Communication in American Engineering Programs 1850–1950.* Norwood: Ablex Publishing Corporation, 1996.

Lannon, John M. *Technical Writing.* 6th ed. New York: HarperCollins, 1994.

Latour, Bruno. *Science in Action: How to Follow Scientists and Engineers through Society.* Cambridge: Harvard University Press, 1987.

Latour, Bruno, and Steve Woolgar. *Laboratory Life: The Construction of Scientific Facts.* Princeton: Princeton University Press, 1986.

Lawrence, Nelda R., and Elizabeth Tebeaux. *Writing Communications in Business and Industry*. 3rd ed. Englewood Cliffs: Prentice-Hall, 1982.

Lawson, Hilary. "Stories about Stories." *Dismantling Truth: Reality in the Post-Modern World*. Ed. Hilary Lawson and Lisa Appignanesi. London: Weidenfeld and Nicolson, 1989. xi–xxviii.

Lay, Mary M. "Feminist Theory and the Redefinition of Technical Communication." *Journal of Business and Technical Communication* 5 (1991): 348–70.

———. "Gender Studies: Implications for the Professional Communication Classroom." *Professional Communication: The Social Perspective*. Ed. Nancy Roundy Blyler and Charlotte Thralls. Newbury Park: Sage, 1993. 215–29.

———. "The Value of Gender Studies to Professional Communication Research." *Journal of Business and Technical Communication* 8.1 (January 1994): 58–90.

Layton, E. "Conditions of Technological Development." *Science, Technology, and Society: A Cross-Disciplinary Perspective*. Ed. Ina Spiegel-Rosing and Derek de Solla Price. London: Sage Publications, 1977. 197–222.

Lebovics, Herman. "The Uses of America in Locke's *Second Treatise of Government*." *Journal of the History of Ideas* 47.4 (1986): 567–81.

Leitch, Vincent B. *Cultural Criticism, Literary Theory, Poststructuralism*. New York: Columbia University Press, 1992.

Lewis, J. Slater. "Works Management for the Maximum of Production." *The Engineering Magazine* 18.1 (October 1899): 59–68.

Locke, John. *An Essay Concerning Human Understanding*. 1693. Amherst: Prometheus Books, 1995.

———. *Several Papers Relating to Money, Interest and Trade, &c.* 1696. New York: Augustus M. Kelley Publishers, 1968.

———. *Two Treatises of Government*. 1690. New York: Hafner Publishing Company. 1956.

Longo, Bernadette. "Teaching Technical Communication with the Community Model: Some Questionable Side Effects." Paper presented at the IEEE Professional Communication Conference, Savannah, September 27–29, 1995.

Lupton, Ellen. *Mechanical Brides: Women and Machines from Home to Office*. New York: Cooper-Hewitt National Museum of Design, Smithsonian Institution, and Princeton Architectural Press, 1993.

Lyotard, Jean-François. *The Differend: Phrases in Dispute*. Trans. Georges Van Den Abbeele. 1983. Minneapolis: University of Minnesota Press, 1988.

Martin, Julian. "Natural Philosophy and its Public Concerns." *Science, Culture and Popular Belief in Renaissance Europe*. Ed. Stephen Pumfrey, Paolo L.

Rossi, and Maurice Slawinski. Manchester: Manchester University Press, 1991. 100–18.

Mathes, J. C., and Dwight W. Stevenson. *Designing Technical Reports: Writing for Audiences in Organizations.* Indianapolis: The Bobbs-Merrill Company, 1976.

McCallum, Daniel C. "Superintendent's Report, March 25, 1856." *The Railroads: The Nation's First Big Businesses.* Ed. Alfred D. Chandler. New York: Harcourt, Brace & World, 1965. 101–07.

McCarthy, Lucille Parkinson, "A Psychiatrist Using *DSM-III*: The Influence of a Charter Document in Psychiatry." *Textual Dynamics of the Professions: Historical and Contemporary Studies of Writing in Professional Communities.* Ed. Charles Bazerman and James Paradis. Madison: University of Wisconsin Press, 1991. 358–78.

McCarthy, Lucille Parkinson, and Joan Page Gerring. "Revising Psychiatry's Charter Document DSM-IV." *Written Communication* 11.2 (April 1994): 147–92.

McDonald, Philip B. *English and Science.* New York: D. Van Nostrand Company, 1929.

Michaels, William Benn. *The Gold Standard and the Logic of Naturalism: American Literature at the Turn of the Century.* Berkeley: University of California Press, 1987.

Miller, Carolyn R. "A Humanistic Rationale for Technical Writing." *College English* 40 (1979): 610–17.

———. "Rhetoric and Community: The Problem of the One and the Many." *Defining the New Rhetorics.* Ed. Theresa Enos and Stuart C. Brown. Newbury Park: Sage, 1993. 79–94.

Miller, Carolyn R., and Jack Selzer. "Special Topics of Argument in Engineering Reports." *Writing in Non-academic Settings.* Ed. Lee Odell and Dixie Goswami. New York: Guilford, 1985.

Mills, Gordon H., and John A. Walter. *Technical Writing.* 3rd ed. New York: Holt, Rinehart and Winston, Inc., 1970.

Mouffe, Chantal. "Democratic Citizenship and the Political Community." *Community at Loose Ends.* Ed. Miami Theory Collective. Minneapolis: University of Minnesota Press, 1991. 70–82.

Moxon, Joseph. *Mechanick Exercises, Or the Doctrine of Handy-Works.* 3rd ed. 1703. New York: Praeger Publishers, 1970.

Mulkay, Michael J. "Sociology of the Scientific Research Community." *Science, Technology, and Society: A Cross-Disciplinary Perspective.* Ed. Ina Spiegel-Rosing and Derek de Solla Price. London: Sage Publications, 1977. 93–148.

Navin, Thomas R. *The Whitin Machine Works Since 1831*. Cambridge: Harvard University Press, 1950.

Nicholson, Joel D., et al. "United States versus Mexican Perceptions of the Impact of the North American Free Trade Agreement." *Journal of Business and Technical Communication* 8.3 (July 1994): 344–52.

Niles, Henry E., Mary Cushing Niles, and James C. Stephens. *The Office Supervisor: His Relations to Persons and to Work*. 3rd ed. New York: John Wiley & Sons, 1959.

Nolan, Robert E., Richard T. Young, and Ben C. DiSylvester. *Improving Productivity Through Advanced Office Controls*. New York: American Management Associations, 1980.

Nystrand, Martin, Stuart Greene, and Jeffrey Wiemelt. "Where Did Composition Studies Come From? An Intellectual History." *Written Communication* 10.3 (July 1993): 267–333.

Olsen, Leslie A., and Thomas N. Huckin. *Technical Writing and Professional Communication*. 2nd ed. New York: McGraw-Hill, 1991.

Painter. F.V.N. *A History of Education*. New York. D. Appleton and Company, 1896.

Paradis, James. "Bacon, Linaeus, and Lavoissier: Early Language Reform in the Sciences." *New Essays in Technical and Scientific Communication: Research, Theory, Practice*. Ed. Paul V. Anderson, R. John Brockman, and Carolyn R. Miller. Farmingdale: Baywood, 1983. 200–24.

———. "Evolution and Ethics in Its Victorian Context." *Evolution and Ethics, with New Essay on Its Victorian and Sociobiological Context*. Thomas H. Huxley. Princeton: Princeton University Press, 1989. 3–56.

———. *T. H. Huxley: Man's Place in Nature*. Lincoln: University of Nebraska Press, 1978.

Paradis, James, David Dobrin, and Richard Miller. "Writing at Exxon ITD: Notes on the Writing Environment of an R&D Organization." *Writing in Nonacademic Settings*. Ed. Lee Odell and Dixie Goswami. New York: Guilford, 1985. 281–307.

Patterson, Robert Alexander. "The Importance of Physics to Engineering." *Engineering Education: Essays for English*. Ed. Ray Palmer Baker. New York: John Wiley & Sons, 1928. 141–56.

Pickering, Andrew. "From Science as Knowledge to Science as Practice." *Science as Practice and Culture*. Ed. Andrew Pickering. Chicago: University of Chicago Press, 1992. 1–28.

Porter, James E. "The Role of Law, Policy, and Ethics in Corporate Composing: Toward a Practical Ethics for Professional Writing." *Professional*

Communication: The Social Perspective. Ed. Nancy Roundy Blyler and Charlotte Thralls. Newbury Park: Sage, 1993. 128–43.

Price, Derek de Solla. "The Relations between Science and Technology and Their Implications for Policy Formation." *Science and Technology Policies.* Eds. Gabor Strasser and Eugene M. Simons. Cambridge: Ballinger Press, 1973.

Pumphrey, Stephen. "The History of Science and the Renaissance Science of History." *Science, Culture and Popular Belief in Renaissance Europe.* Ed. Stephen Pumfrey, Paolo L. Rossi, and Maurice Slawinski. Manchester: Manchester University Press, 1991. 48–70.

Pursell, Carroll. *The Machine in America: A Social History of Technology.* Baltimore: Johns Hopkins University Press, 1995.

Quackenbos, John Duncan. *Practical Rhetoric.* New York: American Book Company, 1896.

Rattansi, P. M. "The Social Interpretation of Science in the Seventeenth Century." *Science and Society 1600–1900.* Ed. Peter Mathias. Cambridge: Cambridge University Press, 1972.

Renard, Georges. *Guilds in the Middle Ages.* 1918. New York: Augustus Kelley Publishing, 1968.

Rev. of *A Sketch of the Life and Works of George W. Whistler, C. E.*, by George L. Vose. *The Railroad and Engineering Journal* 61:2 (February 1887): 52.

Rev. of *Efficiency as a Basis of Operation and Wages*, by Harrington Emerson. *Railroad Age Gazette* 47 (November 5, 1909): 855–56.

Richards, Frank. "Is Anything the Matter with Piece Work?" *Transactions of the American Society of Mechanical Engineers* 25 (1903): 68–92.

Rickard, T. A. *A Guide to Technical Writing.* 2nd ed. San Francisco: The Mining and Scientific Press, 1910.

———. *Man and Metals: A History of Mining in Relation to the Development of Civilization.* 2 vols. New York: McGraw-Hill, 1932.

Rogers, G. A. J. "Introduction." *Locke's Philosophy: Content and Context.* Ed. G. A. J. Rogers. Oxford: Clarendon Press, 1994. 1–28.

Rossi, Paolo L. "Society, Culture and the Dissemination of Learning." *Science, Culture and Popular Belief in Renaissance Europe.* Ed. Stephen Pumfrey, Paolo L. Rossi, and Maurice Slawinski. Manchester: Manchester University Press, 1991. 143–75.

Rubin, Donald. "Introduction: Four Dimensions of Social Construction in Written Communication." *The Social Construction of Written Communication.* Ed. Bennett Rafoth and Donald Rubin. Norwood, N.J.: Ablex, 1988.

Rupert, Mark. *Producing Hegemony: The Politics of Mass Production and American Global Power*. Cambridge: Cambridge University Press, 1995.

Russell, Bertrand. "Where Is Industrialism Going?" *Engineering Education: Essays for English*. 2nd ed. Ed. Ray Palmer Baker. New York: John Wiley & Sons, 1928. 217–33.

Salsbury, Stephen. "The Emergence of an Early Large-Scale Technical System: The American Railroad Network." *The Development of Large Technical Systems*. Ed. Renate Mayntz and Thomas P. Hughes. Frankfurt am Main: Campus Verlag, 1988.

Sauer, Beverly A. "Communicating Risk in a Cross-Cultural Context: A Cross-Cultural Comparison of Rhetorical and Social Understandings in U.S. and British Mine Safety Training Programs." *Journal of Business and Technical Communication* 10.3 (July 1996): 306–29.

Schivelbusch, Wolfgang. *The Railway Journey: Trains and Travel in the 19th Century*. Trans. Anselm Hollo. 1977. New York: Urizen Books, 1979.

Selzer, Jack. "Intertextuality and the Writing Process: An Overview." *Writing in the Workplace: New Research Perspectives*. Ed. Rachel Spilka. Carbondale: Southern Illinois University Press: 1993. 171–80.

Seth, James. *English Philosophers and Schools of Philosophy*. London: J. M. Dent & Sons, 1912.

Shotwell, James Thomson. "Mechanism and Culture." *Engineering Education: Essays for English*. 2nd ed. Ed. Ray Palmer Baker. New York: John Wiley & Sons, 1928. 201–16.

Smart, Graham. "Genre as Community Invention: A Central Banks' Response to Its Executives' Expectations as Readers." *Writing in the Workplace: New Research Perspectives*. Ed. Rachel Spilka. Carbondale: Southern Illinois University Press: 1993. 124–40.

Smith, Terry. *Making the Modern: Industry, Art, and Design in America*. Chicago: University of Chicago Press, 1993.

Spedding, James, Robert Ellis, and Douglas Heath. *The Works of Francis Bacon*. 14 vols. New York: Hurd and Houghton, 1869–1872.

Sprat, Thomas. *History of the Royal Society*. Ed. Jackson I. Cope and Harold Whitmore Jones. 1667. St Louis: Washington University Studies; London: Routledge & Kegan Paul Ltd., 1959.

Stahl, William H. *Roman Science: Origins, Development, and Influence to the Later Middle Ages*. Madison: University of Wisconsin Press, 1962.

Stygall, Gail. "Texts in Oral Context: The 'Transmission' of Jury Instructions in an Indiana Trial." *Textual Dynamics of the Professions: Historical and Contem-*

porary Studies of Writing in Professional Communities. Ed. Charles Bazerman and James Paradis. Madison: University of Wisconsin Press, 1991. 234–53.

Taylor, Frederick W. "A Piece-Rate System, Being a Step Toward Partial Solution of the Labor Problem." *Transactions of the American Society of Mechanical Engineers* 16 (1895): 856–903.

———. "Shop Management." *Scientific Management.* 1911. New York, Harper & Brothers Publishers, 1947.

———. "Testimony Before the Special House Committee." *Scientific Management.* 1911. New York, Harper & Brothers Publishers, 1947.

Taylor Society, The. *Scientific Management in American Industry.* New York: Harper and Brothers Publishers, 1929.

Terry, George R. *Office Management and Control: The Administrative Managing of Information.* 6th ed. Homewood: Richard D. Irwin, Inc., 1970.

Thralls, Charlotte. "Rites and Ceremonials: Corporate Video and the Construction of Social Realities in Modern Organizations." *Journal of Business and Technical Communication* 6 (1992): 381–402.

Thurston, Robert H. "Our Progress in Mechanical Engineering: The President's Annual Address." *Transactions of the American Society of Mechanical Engineers* 2 (1881): 425–53.

Tiles, Mary. *Bachelard: Science and Objectivity.* Cambridge: Cambridge University Press, 1984.

Towne, Henry. "Foreword to 'Shop Management.'" *Scientific Management.* Frederick Winslow Taylor. 1911. New York, Harper & Brothers Publishers, 1947.

———. "Gain-Sharing." *Transactions of the American Society of Mechanical Engineers* 10 (1888): 600–26.

Trimbur, John. "Consensus and Difference in Collaborative Learning." *College English* 51 (1989): 602–16.

Tully, John. "Rediscovering America: The *Two Treatises* and Aboriginal Rights." *Locke's Philosophy: Content and Context.* Ed. G. A. J. Rogers. Oxford: Clarendon Press, 1994. 165–96.

Urwick, L., and E. F. L. Brech. *The Making of Scientific Management.* 2 vols. London: Sir Isaac Pitman and Sons Ltd., 1966.

Wagner, David L. "The Seven Liberal Arts and Classical Scholarship." *The Seven Liberal Arts and the Middle Ages.* Ed. David L. Wagner. Bloomington: Indiana University Press, 1983. 1–31.

Wahlstrom, Billie. "Designing a Research Program in Scientific and Technical Communication: Setting Standards and Defining the Agenda." *Foundations for Teaching Technical Communication: Theory, Practice, and Program Design.* Ed. Katherine Staples and Cezar Ornatowski. Greenwich: Albex Publishing Corporation, 1997. 299–316.

Wellborn, G. P., L. B. Green, and K. A. Nall. *Technical Writing.* Boston: Houghton Mifflin Company, 1960.

Wendell, Barrett. *English Composition: Eight Lectures Given at the Lowell Institute.* New York: Charles Scribner's Sons, 1891.

Whewell, William. *The Philosophy of the Inductive Sciences. Founded upon Their History.* 1847. New York: Johnson Reprint Company, 1967.

Williams, Cecil B., and E. Glenn Griffin. *Effective Business Communication.* 3rd ed. New York: The Ronald Press Co., 1966.

Williams, Raymond. *Keywords: A Vocabulary of Culture and Society.* Rev. ed. New York: Oxford University Press, 1983.

Williamson, George. *The Senecan Amble: A Study of Prose Form from Bacon to Collier.* 1951. Chicago: Phoenix Books–University of Chicago Press, 1966.

Winterowd, W. Ross. "Literacy, Linguistics, and Rhetoric." *Teaching Composition: Twelve Bibliographic Essays.* Ed. Gary Tate. Fort Worth: Texas Christian University Press, 1987. 265–90.

Wrege, Charles D., and Ronald G. Greenwood. *Frederick W. Taylor the Father of Scientific Management: Myth and Reality.* Homewood: Business One Irwin, 1991.

Yates, Frances A. *Giordano Bruno and the Hermetic Tradition.* Chicago: University of Chicago Press, 1964.

Zak, Michele Wender. "'It's Like a Prison in There': Organizational Fragmentation in a Demographically Diversified Workplace." *Journal of Business and Technical Communication* 8.3 (July 1994): 281–98.

Zappen, James. "Bacon, Francis." *Encyclopedia of Rhetoric and Composition: Communication from Ancient Times to the Information Age.* Ed. Theresa Enos. New York: Garland Publishing, Inc., 1996. 61–63.

———. "Francis Bacon and the Historiography of Scientific Rhetoric." *Rhetoric Review* 8.1 (Fall 1989): 74–90.

———. "Francis Bacon on Democratic Science and Plain Prose." Paper presented at the Speech Communication Association Annual Meeting, Miami Beach, November 18–21, 1993.

Index